PESTICIDE IMPACT ON STREAM FAUNA

WITH SPECIAL REFERENCE TO

MACROINVERTEBRATES

PESTICIDE IMPACT ON STREAM FAUNA
WITH SPECIAL REFERENCE TO
MACROINVERTEBRATES

R. C. MUIRHEAD-THOMSON

Lately Research Fellow
Royal Holloway College,
University of London

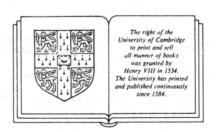

The right of the
University of Cambridge
to print and sell
all manner of books
was granted by
Henry VIII in 1534.
The University has printed
and published continuously
since 1584.

CAMBRIDGE UNIVERSITY PRESS

CAMBRIDGE
NEW YORK   PORT CHESTER   MELBOURNE   SYDNEY

CAMBRIDGE UNIVERSITY PRESS
Cambridge, New York, Melbourne, Madrid, Cape Town, Singapore,
São Paulo, Delhi, Dubai, Tokyo

Cambridge University Press
The Edinburgh Building, Cambridge CB2 8RU, UK

Published in the United States of America by Cambridge University Press, New York

www.cambridge.org
Information on this title: www.cambridge.org/9780521125284

First published 1987
Reprinted 1989
This digitally printed version 2009

A catalogue record for this publication is available from the British Library

Library of Congress Cataloguing in Publication data
Muirhead-Thomson, R. C.
Pesticide impact on stream fauna.
Bibliography
Includes index.
1. Pesticides – Environmental aspects. 2. Freshwater
invertebrates – Ecology. 3. Stream fauna – Ecology.
I. Title.
QH545.P4M85 1987 591.2′4 86–20696

ISBN 978-0-521-30967-7 Hardback
ISBN 978-0-521-12528-4 Paperback

CONTENTS

PREFACE AND ACKNOWLEDGEMENTS

The preparation of this review, and the final writing up, was only made possible by generous support from two organisations, namely the Water Research Centre, Medmenham, Marlow, Buckinghamshire, UK, and Jealotts Hill Research Station (ICI Plant Protection), Bracknell, Berkshire. I am greatly indebted to the sponsors concerned, Mr J. F. Solbe of the WRC and Dr B. G. Johnen of ICI for authorising donations without which this project would not have seen the light of day. Their assistance was particularly vital, coming as it did at a period when research funding in general, and in this field in particular, was being severely restricted in this country.

I am also grateful for the cooperation of many colleagues through correspondence from overseas, who have not only been kind enough to send reports and material not easily available here, but who have also – by correspondence – provided up-to-date progress information and comments. I am particularly indebted to the following: Dr Peter Kingsbury of the Canadian Forest Service; Dr D. C. Eidt of the Maritime Forest Research Centre, New Brunswick; Dr P. E. K. Symons of the Fisheries Research Board, Canada; Dr W. O. Haufe of the Animal Parasitology Research Station, Lethbridge, Alberta; Dr Joan Trial of the Department of Zoology, University of Maine, US; Prof. John Giesy of the Pesticide Research Centre, Michigan State University, US; Dr. Aarne Lamsa, of the Great Lakes Fishery Commission, Ann Arbor, Michigan; Dr L. A. Norris of the USDA Forest Service, Oregon State University, US; Dr Dan Kurtak of the OCP (Onchocerciasis Control Programme), Ouagadougou, Burkina Faso (Upper Volta); Dr C. Dejoux of ORSTOM (OCP, Burkina Faso); Dr J. M. Elouard of ORSTOM (OCP); Dr J. W. Everts, Dept. of Toxicology, Wageningen, Netherlands, and Dr Peter Matthiessen, of the Centre for Overseas Pest Research, London.

Below are listed all of the abbreviations used in this volume with the page of their first occurrences

ABBREV	CHAPTER	PAGE
AI	3	23
BHC (HCH)	2	11
DDD	2	13

I

INTRODUCTION AND ORIGIN
OF PESTICIDES IN
RUNNING WATERS

ONE

INTRODUCTION

At the time when I last carried out a review of the subject of pesticides and freshwater fauna (Muirhead-Thomson, 1971) it was still possible for a single author to do justice, within one book, to the information then available regarding all forms of freshwater animal life and all types of freshwater body. In the 15 years since that book was published, there has not only been an enormous proliferation of knowledge about this subject but also noteworthy changes in emphasis and priorities. There has been increasing specialisation within this general subject as well, making it increasingly difficult for a single author to encompass all aspects of this problem. For all these reasons, the scope of the present review is restricted to running waters, rivers and streams, and to the macroinvertebrate fauna of such water bodies. The restriction to macroinvertebrate fauna is dictated in part by the fact that a great deal of the voluminous literature in the last 15 years deals with studies on the reactions of freshwater fish, to such an extent that a competent review of that aspect, including all the physiological work on uptake and retention of pesticides by different organs, would require a separate volume. However, one aspect of those fish studies cannot be omitted from any review devoted to aquatic macroinvertebrates, that is the effect of pesticides and allied toxic chemicals on feeding habits of fish in so far as these are influenced by drastic changes in the availability of different invertebrate fish food organisms, as measured by changes in the composition of the stomach contents.

The restriction of the present review to running water habitats is dictated by the simple fact that, within that period, the most noticeable advances have been concerned with pesticide impact on river and stream ecology. The great international advances within that field have made it imperative to review and discuss that work in depth; accordingly, limitations of space have made it unrealistic to attempt a similar review of advances in lakes, ponds and other static water bodies and their fauna.

Although this may seem a drastic limitation of subject, there are other reasons – besides limitations of space – to justify the choice of subject. Firstly, the problems posed by pesticide contamination of running waters are very different from those encountered in the static waters of lakes, ponds and reservoirs (Hynes, 1972). Prominent among these differences is the question of choice (or design) of basic capture and sampling techniques essential in

3

evaluating the reactions of running water macroinvertebrates to pesticide contamination. The considerable advances in the development and use of these evaluation techniques, and in the interpretation of capture data in pesticide studies, have found parallels within this same period in the field of pure freshwater biology, where there have also been noteworthy developments in drift studies and in sampling the bottom fauna of running waters.

Secondly, the last 15 years have seen a change in emphasis and priorities in pesticide contamination of freshwater bodies. Up till the end of the 1960s many of these studies were concerned with the impact of DDT and its allied chlorinated hydrocarbon insecticides, and with the capacity of these chemicals to accumulate in animal tissues leading to progressive concentration in the ecological food chain. As a result of these undesirable tendencies to biological accumulation – which, in some lakes, could lead to the persistence of toxic residues for several years – DDT and its close allies (dieldrin and aldrin) have been gradually phased out from pest control projects or have been banned completely in some countries.

The less persistent chlorinated hydrocarbons such as endosulphan (Thiodan), methoxychlor and toxaphene are still being used, but in general the last 15 years have witnessed the greatly increased use of the less persistent insecticides belonging to three different groups, organophosphorus compounds, carbamates and the pyrethrins/pyrethroids. All three groups in contrast to DDT and dieldrin, are characterised by more rapid 'biodegradability' in the environment and limited capacity to accumulate in aquatic food chains. These are more desirable properties from the point of view of environmental contamination and the protection of 'non-target' organisms, especially when combined with low mammalian toxicity. Against this is the fact that these new insecticides, though not necessarily other pesticides, can be highly toxic to insects and arthropods in general, which form the bulk of freshwater macroinvertebrates. With these points in mind, it can perhaps be more readily appreciated that running water studies in pesticide ecology have reached a stage where they merit a full and separate review.

Earlier in this chapter, attention was drawn to the development of sampling methods best suited to evaluate the impact of pesticides on the invertebrate populations of running water. As the validity of these various capture and netting procedures is crucial to a better understanding of both the immediate and the long-term effects of pesticide pressure, a great deal of space will be devoted to an assessment of progress made in recent years by different groups of workers, often making quite independent approaches to common basic problems. Because of varying conditions and varying needs in different situations, the application of these sampling techniques is still in an experimental and exploratory stage. The time has not yet come when it is possible to lay down guidelines or recommend standard procedures for running water macroinvertebrates. This is in marked contrast to the long-accepted standard procedures used in fish toxicity studies. Evidence will be produced in the forthcoming chapters to show that whatever advantages there may be in laying down internationally acceptable standard procedures and standard criteria as early as possible in this rapidly developing discipline, these may be more than offset by the fact that such rigid and arbitrary standards leave little room for the flexible approach essential for a complete understanding of the organism's reactions to its contaminated environment.

For these reasons, it is not the intention in this book to provide a manual of instruction, laying down guidelines for field and laboratory procedures, but rather to discuss and assess the range of experience recorded by investigators working in different countries, often under very different conditions of climate, topography and faunal composition. It is felt that a full understanding of the different evaluation practices and criteria can only be reached if the background to each project is appreciated. This is particularly applicable to evaluation methods used in the field, where it is felt that a description of the varied methodologies in a single chapter – but out of context – would be a confusing introduction to the subject. Accordingly, the practice adopted has been to describe the different evaluation methods used, and the results obtained, in the course of a general description of each main investigation reviewed. Only after these practices have been seen against the background of the particular problems encountered in each environmental study and the particular objectives of each study, can they all be reviewed and assessed in a final chapter.

The period under review has been marked by rapid developments in the experimental approach to problems of evaluation and the development of controlled laboratory techniques specially designed for the specific require-ments of running water macroinvertebrates. The successful maintenance or culture of different stream biota under laboratory conditions is a subject which has for too long been neglected by freshwater biologists themselves. Accordingly, pesticide ecologists have had to devise appropriate techniques *ab initio*. From the progress made so far, it is clear that a laboratory phase of evaluation must play an increasingly important part in the overall strategy of evaluating pesticide impact. Continued improvement in the design and application of laboratory simulated-streams with the ultimate objective of establishing different stream macroinvertebrates in a controlled running water environment as closely akin to the natural habitat as possible, offers a great hope for the future and a real challenge to the ingenuity of future pesticide ecologists.

Finally, studies on pesticide impact in recent years have been greatly assisted by technical advances in physicochemical methods for determining pesticide residues in various components of the aquatic environment, inclu-ding vegetation, silt and water, as well as in the aquatic organisms themselves. Many widely used pesticides can now be detected in water at levels as low as

0.1 µg/l 0.1 parts per billion). These technical advances have resulted in a great proliferation of published works on this aspect, and there are now several journals devoted exclusively to pesticide monitoring and pesticide residues. So much so that, in many respects, knowledge about the occurrence and distribution of pesticides in the environment, including streams and rivers, has far outrun any knowledge about the ecological consequences of such pesticide presence. One of the major contributions of those advances has been to provide firm confirmation of the fact that in running waters exposed to pesticide treatment, either directly or indirectly, the chemical tends to disappear fairly rapidly – usually within a few days after application. Equally important is the fact that residue studies now play an essential role in most studies on experimental applications of pesticides to water. This work has provided new and vital knowledge about the patterns of pesticide distribution in running waters in relation to applied dosage and distance downstream from the application point. Much more information is now available about the precise concentrations of toxicants to which stream organisms are exposed, and the duration of this exposure. This knowledge is also being increasingly used in laboratory evaluation techniques and in studies on experimental water plots, in which various factors such as adsorption onto surfaces or differential dispersion of the toxic chemical in the water column, may produce significant differences between the known applied concentration and the real concentration in the water body, both in space and time.

The aim of this book is not so much to summarise or catalogue information available from published reports, but rather to examine the great variety of methods which have been used to provide this basic knowledge. Consequently, there has been special emphasis on those studies in which problems of evaluation have been examined critically and experimentally. In selecting such material from a multitude of journals and reports – not all of which are easily available – it is perhaps inevitable that one or two important contributions to the subject have not received due recognition, or have been overlooked completely. If this does happen, the reviewer must offer his sincere apologies and regrets, and hope that such omissions are not interpreted in any way as disparagement of that particular work, but must be put down to human error and the increasing limitations of a single author to cover such a vast field.

One point which needs to be clarified at this early stage is the use of different notations to express the concentration of pesticide in the environment in general and in water in particular. These are summarised as follows:

parts per million, ppm, *or* milligrams per litre   (mg/L *or* mg/l *or* mg/l^{-1});

parts per billion (thousand million), ppb, *or* micrograms per litre   µg/L *or* µg/l, *or* µg/l^{-1};

parts per thousand billion   ng/L.

All these terms have been used by authors quoted in this review – sometimes different notations by the same author in different publications. However, the terms quoted in the text have been standardised (as mg/l, µg/l or ng/l) in order to facilitate comparisons.

TWO

ORIGINS OF PESTICIDES IN

RUNNING WATERS

The term 'pesticide' embraces a wide range of toxic chemicals used for controlling or eradicating undesirable forms of life. Compounds specifically designed for the control of insects and other arthropods, i.e. insecticides, make up the bulk of these; another range of pesticides is designed to deal with undesirable fish (both predatory and competitive) while still others were developed for use against the aquatic snails which harbour intermediate stages of human parasites. The term pesticide now conveniently includes herbicides, chemicals specifically designed for control of undesirable plant growth, and this inclusion recognises the fact that the greatly increased use of herbicides in recent years has, in many cases, placed these chemicals equal to or ahead of insecticides as major environmental contaminants (Balk & Koeman, 1984)

CONTAMINATION AS A DIRECT CONSEQUENCE
OF PEST CONTROL OPERATIONS

DIRECT APPLICATION OF PESTICIDE TO WATERBODY

Pesticide contamination of running waters can occur in many different ways and from many different sources, and may be only of short duration or it may be prolonged. In view of the emphasis in this review on the problems of evaluation, it would be convenient to consider pesticide contamination under two main categories. In the first category would be listed all those cases where the presence of pesticide in running water is the direct consequence of control operations carried out against undesirable fauna or flora. In some cases, the application of pesticide is made directly to the stream or river as, for example: in controlling the aquatic larvae of the biting black fly, *Simulium*; for controlling 'trash' and predatory fish, or predatory fish larvae; for control of nuisance insects of aquatic origin; and for the removal of undesirable aquatic weeds by applications of herbicides.

8

Origins of pesticides in running waters

Whether the chemical is applied from the ground, e.g. from boats moored across a large river, or by spraying from fixed-wing aircraft or from helicopter, the common factor is that the running water is the direct target of the operation.

AERIAL SPRAYING AGAINST TERRESTRIAL PESTS

Pesticides may contaminate a stream or river in the course of spray operations directed against terrestrial pests, mainly those of concern to agriculture and forestry but also including medically important insects. Whether these operations are directed against forest pests such as the spruce budworm in Canada and the US, or against the resting sites of tsetse flies in the thick gallery forest fringing rivers in West Africa, the underlying water bodies are particularly vulnerable to immediate contamination from such spraying operations.

The common factor in all these different situations arising from pest control operations is that the identity of the contaminating chemical is known, as also is its timing, duration, rate of application, and the total area under treatment. In all these cases the subsequent evaluation of the impact on stream fauna, or flora, can be related to a clearly defined episode in which chemical is applied directly to the water body at known concentration (in terms of parts per million), or to the environment of the stream (in terms of grams per hectare or pounds per acre). More precise physicochemical monitoring of the affected water, in relation to exact timing and duration of impact, can then be carried out in order to build up a more solid basis of evaluation.

The great wealth of information produced by the many and varied field investigations into situations of this kind has indeed provided the bulk of material reviewed in this book, and has been a major contribution to the great advances in knowledge within the last 10 years.

PESTICIDE APPLICATION TO ADJACENT
STATIC WATERS

Before leaving this discussion of contamination due to pest control operations, it is important to know that static water bodies such as lakes, ponds, marshes and rice fields are also subject to direct pesticide application for control of mosquitoes, midge larvae, rice pests, undesirable fish and aquatic weeds, as well as aquatic snails.

While such static water bodies are, in general, outside the scope of this review, various running water bodies such as irrigation and drainage

9

channels, outlet streams etc. associated with these mainly static water bodies could be liable to direct pesticide contamination in the course of these spraying operations. This has perhaps particular relevance to all the flowing water associated with irrigation and drainage in wet rice cultivation, for example in Indonesia where endosulphan was applied on an extensive scale for control of rice stem borer (Gorbach *et al.*, 1971 *a*,*b*). The monitoring of these pesticide residues is important in connection with their likely effect on sensitive food fish and on the fish-food organisms. A similar situation involving fish culture and wet rice cultivation also exists in Malaysia where concern has been expressed about the continued use of persistent and toxic pesticides in rice fields, and about the confirmed presence of such residues in the water (Meier, Fook & Lagler, 1983). Attention is drawn to these examples of static waters, and many others which could be quoted, mainly to emphasise that in the evaluation of such environmental situations the common factor which would include them in the category of contamination due to pest control operations is the known identity of the pesticide, and the exact timing, duration, frequency and application rates used in the spraying operations.

CONTAMINATION NOT CLEARLY ATTRIBUTABLE
TO LOCAL PEST CONTROL OPERATIONS

The second category of pesticide contamination of running waters includes a wide range of situations which are not the immediate result of planned or routine application of pesticide. At one extreme are the sporadic and usually unpredictable cases of sudden high pesticide contamination due to spillage of concentrates, or careless disposal or washing of spray equipment in or near streams. By the time the spillage has come to the attention of the appropriate authorities – perhaps through reports of local fish kills – the pesticide has been washed downstream, and the water is running clean and pure again. Only if these conditions are re-created experimentally, is there any real chance of evaluating their environmental impact.

At the other extreme are the very low, but persistent, pesticide residues recorded in rivers and streams in practically all areas of the world where pesticide has been used. Many different chemicals may be present in these residues, but in general the dominant role has been played by the persistent chlorinated hydrocarbon insecticides, DDT, dieldrin and aldrin, which may still be detectable several years after their use has been discontinued in a particular area. Between these two extremes, this second category included many different types of pesticide contamination from industrial sources, often on an unpredictable and intermittent pattern, the evaluation of which is further complicated by the presence of other industrial contaminants (such as ammonia, cyanides, heavy metals etc.).

CONTAMINATION BY ACCIDENT OR SPILLAGE

The contamination of running waters through spillage, careless disposal of surpluses, or accident is liable to occur in any country where pesticide is used on a regular basis. In countries such as the UK, where streams and rivers are generally free from major pesticide pollution (Royal Commission, 1979), such incidents take on an added significance, and may account for a significant proportion – up to 25% in some regions – of reported fish kills (Holden, 1972). The main interest in the continuing occurrence of such incidents, especially in Scotland (Bowen, 1982) is that although their impact on stream invertebrates and stream ecology appears to be transitory, the very nature of the incident makes precise evaluation difficult. In addition, contamination may be insidious, or from a distant source. This was well illustrated in the Chew Lake, an impounded reservoir of the Bristol Waterworks Co. in which the condition of the water, which had remained stable from 1960 to 1967, deteriorated in 1968 due to lack of grazing microcrustacean *Cladocera*. The cause was eventually traced to contamination by pesticide from sheep dips passing underground to an egress in a feeder stream (Bays, 1969).

Only the experimental replication of various degrees of spillage can provide essential data on the likely impact of such occurrences. The value of such a controlled application of sheep dip insecticide – in this case gamma BHC – was amply demonstrated over 25 years ago (Hynes, 1961).

CONTAMINATION BY PESTICIDES FROM INDUSTRIAL SOURCES

Pesticide contamination of streams and rivers by industrial effluents is well exemplified by the incorporation of insecticides in the manufacture of textiles. Studies carried out in the US on a case of river contamination from a woollen mill in which dieldrin was used for moth-proofing, have thrown a great deal of light on the reactions of invertebrates to such situations (Wallace & Brady, 1974). Dieldrin residues in the larvae of *Simulium vittatum* showed a dramatic increase which varied from 56-fold in April to 1178-fold in February. The highest residue in any animal tested was 90–103 ppm in larvae of the trichopteran *Cheumatopsyche* while, in contrast, the hellgramite, *Corydalus cornuta* had the lowest amount of dieldrin. That study also revealed consid-

11

erable differences in uptake of chemical on the part of two closely related trichopterans, *Cheumatopsyche* and *Hydropsyche*, possibly associated with different feeding sites or habits.

Similar situations caused by dieldrin-containing discharges from textile mills have existed in several areas in the UK over the last 20–25 years. High concentrations of dieldrin were found in the River Aire and the River Calder which flow through areas associated with textile proofing in Halifax, Huddersfield and Bradford, while in the same year – 1966 – major contamination in the upper part of the Lee was found to be gamma BHC (Lowden, Saunders & Edwards, 1969). More recently, dieldrin concentrations high enough to have a biological effect were found in the River Holme and some of its tributaries (Brown, Bellinger & Day, 1979). Dieldrin, like other organic pesticides including Eulan WA New (which is largely replacing it), is not significantly removed from the effluent by most conventional sewage treatment works.

With the recession in the textile industry, and the introduction of alternative chemicals for moth-proofing, use of dieldrin in the UK has been mainly restricted to southwest Scotland and West Yorkshire (Boryslawkyj & Garood, 1985), and one of these residue studies in the Huddersfield area of West Yorkshire has revealed unexpected patterns of dieldrin concentration in the river water affected, patterns which have a considerable bearing on wider problems of pesticide evaluation under similar conditions. The general pattern regarding progressive decrease in concentration of dieldrin from the point of discharge downstream was found to accord with previous records (Brown *et al.*, 1979) in showing that at a point 8 km downstream there were only slight changes in concentration. Nearer the source, regular sampling at hourly intervals throughout the day revealed the existence of large and rapid fluctuations in concentration. These massive changes, sometimes of the order of 500%, could occur within an hour at a time when there was little or no perceptible change in discharge, and no sudden rainfall to account for the changes.

The mean concentration of dieldrin in the affected river was found to be 197 ng/l, almost twice the mean value known to produce sublethal effects on fish. During the period of maximum short-term fluctuations, the mean concentration below the discharge point was 1774 ng/l, (i.e. 1.77 ppb). These figures appear in truer perspective when considered in the light of the European Commission's finding that surface waters in the European Community would be unlikely to exceed a mean value of 100 ng/l Butijn & Koeman, 1977).

This example illustrates very well the extreme difficulty likely to be encountered in trying to equate changes in stream fauna to time/concentrations of the contaminant, even when the identity of the pesticide chemical is clearly established, and its point of discharge into the river pinpointed. A record of mean concentration of dieldrin in the river as a whole would fail completely to take into account the occurrence of short pulses of very high concentration which might produce an entirely different range of reactions in the river fauna.

PERSISTENCE OF ORGANOCHLORINE RESIDUES
IN RUNNING WATER

While there has been a steady falling off in the agricultural use of dieldrin and DDT in Great Britain in recent years, organochlorine residues continue to be found in surface waters. This is partly a reflection of the very persistent nature of the metabolites of DDT, such as DDD and DDE, and partly a consequence of continued domestic use of organochlorines, lindane in particular. Samples of sewage sludge from 40 different sewage treatment works in the UK have shown the presence of these organochlorines, lindane (y–HCH), dieldrin and DDE (McIntyre & Lester, 1982) as well as polychlorinated biphenyls (PCBs). The latter contaminants came into considerable prominence in the '70's because of their increasing industrial sources, and their presence in such English rivers as the Avon and Frome (Kpekata, 1975). Since that time the use of PCBs has declined appreciably, and the industrial use has been severely limited to 'closed system' application (McIntyre & Lester, 1982).

The application of such sewage sludges to agricultural land, which plays an important role in economic disposal, could provide the means of contaminating running water bodies at low but continuous levels. Apart from awareness of this situation, and the regular chemical monitoring of rivers for organochlorine residues by River Authorities, little is known about the ecological significance of these extremely low levels of pesticide contamination.

SIGNIFICANCE OF VERY LOW PESTICIDE LEVELS

A pointer to the likelihood that the enviromental significance of very low pesticide concentrations should not be dismissed as negligible comes from an unusual investigation in Northern Ireland (Harper, Smith & Gotto, 1977). Samples of water from the major rivers flowing into Lough Neagh showed that the majority of the samples contained y-BHC ($=$y-HCH) residues at low concentrations of from 10 to 23 ng/l. These concentrations showed a high correlation with high urban population density in the catchment area, and it was evident that the BHC originated in the sewage effluents of domestic origin – probably arising from the use of wood preservatives in the home – and not from agricultural or industrial sources.

The effects of a range of such low concentrations of y-BHC on the survival time of two species of mayfly nymphs, *Baetis rhodani* and *Caenis moesta*, were measured in the laboratory, the basis of comparison being the time taken for 50% of the test animals to die at each concentration. The results showed that major reductions were recorded in the survival times of both species at y-BHC levels as low as 100 ng/l(0.1 ppb), and even slight effects at a concentration of 10 ng/l(0.01 ppb). While the validity of these particular tests has to be assessed in the light of present knowledge about laboratory evaluation techniques (see Chapter 3), there is sufficient material to emphasise the need for further critical study.

CONTAMINATION BY DRAINAGE FROM AGRICULTURAL OR FOREST LAND

Within the second main category of sources of pesticide contamination, i.e. contamination of running waters which is not the immediate and obvious consequence of spraying operations to the water itself or to its immediate environs, the main contribution is undoubtedly from the drainage, run-off and leaching from agricultural or forestry lands regularly treated with pesticide, or with a recent history of such treatment. In countries like the UK where aerial spraying for control of crop pests is still on a relatively small scale, this is probably a negligible contribution to pesticide presence in running waters (except for unusual cases). In the US and Canada, however, with a long history of insecticide usage and its aerial application on a vast scale, these varied sources are sufficient to account for the persistent presence of pesticide residues in running waters. Systematic monitoring of river waters has been practised since the early 1960s in the US (Lichtenberg *et al.*, 1970; Eichelberger & Lichtenberg, 1971) a period when the residual insecticides, DDT, dieldrin and aldrin were being applied on a maximum scale.

In Canada, particular attention has been given to the regular monitoring of pesticide residues in the vast catchment area of the Great Lakes, and its innumerable feeder tributaries (Johnson & Ball, 1972). DDT and dieldrin (DLN) residues were early detected by monitoring in all species of fish in the Great Lakes, and concentrations were sufficient to produce significantly high mortalities among fry of the Coho salmon in hatcheries. At that period, before the phasing out of DDT as a major insecticide, the main contaminating source was from operations for control of Dutch Elm disease with DDT, as well as from mosquito control operations. At that time, substantial quantities of DDT were still entering Lake Michigan via its tributary streams, which may have accounted for the fact that residues in that lake were 2–5 times higher than in Lake Superior (Zabik, 1969; American Chemical Society, 1972).

The greatest transfer of DDT was found to be from a resort area, the Muskoko Lakes, where DDT was used until 1966 for control of biting flies (Miles & Harris, 1971). By considering the concentration of total DDT in the water together with the water flow data for rivers draining this area, the weekly transport of DDT by the rivers was calculated. The peak figure recorded in any 1 week was 11.8 lb (5.36 kg) total DDT, at which time the ratio of (total DDT in fish):(total DDT in water) gave figures up to 1 million, dramatic proof of the bioaccumulation properties of this insecticide. Fish from the resort area itself were found to contain up to 19 ppm total DDT, compared to a high 1.3 ppm in fish from streams draining mainly from agricultural land.

Perhaps even more serious environmentally was the finding that the Muskoko River was still the biggest contributor of DDT in the 1971 samples, 5 years after the use of DDT had officially ended.

A very extensive literature has arisen around these US and Canadian monitoring programmes, and for the moment only a recent and representative example can be examined in any detail. This is the latest in a long series of surveys on agriculture and water quality in the Canadian Great Lakes Basin (Miles & Harris, 1973; Harris & Miles, 1975; Miles, 1976) and deals with pesticide use in 11 agricultural watersheds in relation to its presence in stream water (Frank et al, 1982).

Within the period covered by that survey there have been great changes in pesticide use in Canada. In the late 1960s, organochlorine insecticides predominated, but between 1969 and 1972 almost all persistent members of that group were legislatively eliminated and replaced by the less persistent, biodegradable organophosphorus and carbamate insecticides. At the same time, herbicide use increased until it accounted for more than half the total volume of pesticide. The water samples revealed the presence of 28 separate pesticide compounds, and included insecticides both from current use and from past use. Spills – associated with the process of mixing pesticide and cleaning equipment – and storm run-off from agricultural land accounted for the bulk of stream contamination. Some pesticides, such as the herbicide atrazine and the insecticide endosulphan persisted in soil from one season to the next, and made their contribution to the stream through leaching, storm run-off and internal soil drainage. One of the most significant and disturbing findings was the detection of residues in the water of the persistent organo-chlorines DDT, TDE and HEOD (the active principle of dieldrin), even though none of these compounds had been in current use for several years. In the case of DDT, 41% of the samples exceeded the maximum acceptable level of 3.0 ng/l(0.003 ppb) established as a criterion by the International Joint Commission (IJC).

CONTAMINATION BY AUTUMN-SHED LEAVES

While leaching of soils and run-off from areas with a history of spraying operations probably account for the bulk of continuous low-grade contamination of running waters, there is another significant source associated with forest areas, namely from autumn-shed leaves (Derr, 1974). It is well established that autumn-shed leaves provide energy sources in streams and constitute a major source of food for many stream macroinvertebrates (Kaushik & Hynes, 1968; Anderson & Sedell, 1979). In areas where methoxychlor has replaced DDT, foliage contaminated with this could provide a prolonged source of the insecticide. By means of controlled laboratory experiments, precise dosages of methoxychlor were applied to leaves of different trees from protected areas which were then placed in continuous flow aquaria. Leaves were removed for analysis at specified time intervals of 0, 5, 10, 15 and 20 days; the residues showed that the uptake and retention varied according to the species of tree. On beech, methoxychlor had a half-life of 7 days, while on oak and maple that period extended to 24 and 26 days respectively. Continued leaf fall over a period of several weeks or months could provide a continuing source of contamination long after spraying operations had terminated, and possibly at considerable distances downstream from treated areas.

The extent and persistence of pesticide residues in river systems in Canada and the US is understandably a reflection of the extensive usage of insecticides in agriculture and forestry over a period of many years. It is also a reflection of the great amount of study which has been devoted to this problem in those countries. There is increasing evidence that similar situations exist in other parts of the world where there is intensive pesticide usage, but where the scarcity of scientific data has made it difficult to define the nature of the problem. This is well illustrated in the River Nile where studies have been carried out both in Lake Nubia – the Sudanese portion of Lake Nasser (Elzorgani, 1976; Elzorgani, Abdulla & Ali, 1979) – and at different points on the Egyptian Nile from Aswan to the Delta canals (Aly & Badawi, 1984). Using fish as sensitive measures of low levels of pesticide contamination because of the ability of these chemical residues to concentrate in certain tissues, samples revealed the presence of DDT and its metabolites (DDE, DDD) as well as other organochlorines lindane (BHC) and endrin. There is extensive use of pesticide throughout the agricultural areas of the Nile, but probably one of the major centres of contamination is the cotton-growing region along the Blue and White Niles in central Sudan where organochlorines, DDT, toxaphene and endosulphan have been used on an increasing scale.

16

CONTAMINATION BY AERIAL TRANSPORT

Pesticides may also contaminate running water bodies by means of such natural forms of aerial transport as wind and rain (Hanuska, 1971; Sodergren, Svensson & Ulfstrand, 1972; Sodergren, 1973). It is highly likely for example that atmospheric drift and fallout has made a significant contribution to the organochlorine contamination of Lake Michigan, with a surface area of 22 400 square miles (5 801 600 ha) and an annual rainfall of 33 inches (84 cm) (Johnson & Ball, 1972). Even more direct proof of wind-borne contamination was provided by investigations in northern Pennsylvania prior to the introduction of a large-scale operation for control, with DDT, of an extensive outbreak of Fall Canker Worm over an area of 100 000 acres (40500/ha) of forest land. There had been no previous treatment with insecticide in this region; nevertheless, pre-treatment residue studies revealed the presence of small quantities of DDT and dieldrin in watershed soils and water, the latter at concentrations of 0.0052 ppm. One hour after aerially spraying the forest at the rate of 0.5 lb per acre (561 g/ha) the concentration of DDT in water had risen to 0.024 ppm, and it was concluded that pre-treatment levels must have been the result of aerial transport from distant sources (Cole, Barry & Frear, 1967).

In the course of aerial spraying operations on clearly defined target areas of forest, drift to untreated areas outside that zone may be considerable, up to at least 4 km within 2 h being experienced in control operations against spruce budworm in New Brunswick (Eidt & Sundaram, 1975). More precise studies on aerial transport have been incorporated in several recent studies on the environmental impact of aerial spraying, in which drift has been accurately mapped by means of sample cards or aluminium plates for trapping deposits (Crossland, Shires & Bennett, 1982; Shires & Bennett, 1985).

In operations against tsetse flies in Botswana for example, (Andrews *et al.*, 1983) the insecticide in question, endosulphan, was recovered up to 12.5 km downwind of operations at concentrations sufficient to kill tsetse, the insecticide still being detectable 30 km downwind. In experimental spray operations with permethrin in Quebec, sample cards set on stakes above the surface of forest streams showed significant deposits up to 3 km downstream from the treated area (Kreutzweiser, 1982).

CONTAMINATION BY MORE THAN ONE
TOXIC CHEMICAL

As already noted, the main feature common to all these varied examples of pesticide contamination included in this second category (stream contamination not obviously the consequence of current spraying operations) is that while one particular pesticide may be the dominant contaminant, the chances are that other pesticide contaminants are also present. This makes it increasingly difficult to attribute any observed faunal changes to a particular chemical pesticide, especially at low but continuous concentrations. This difficulty is further aggravated by the fact that other non-pesticide toxicants may also be present, which either individually or in combination produce a different range of effects on the aquatic macroinvertebrate fauna. Such a situation has long been recognised in such large rivers as the Rhine flowing through highly industrial areas, where pesticides detected in the water, such as endosulphan and hexachlorocyclohexane, form only part of a complex of chemical contaminants (Herzel, 1973). In the rivers Rhine and Meuse hundreds of potentially hazardous chemical compounds have been identified (Meijers & van der Leer, 1976; Sloof *et al*, 1983*b*). There is now an urgent need to develop toxicity tests which will enable authorities to predict safe concentrations of these chemicals to aquatic life. Under pressure from the main coordinating authority involved, the European Commission, there has been a much-needed stimulus to scientific progress in this field which in due course will greatly assist in evaluating wider problems of pesticide impact (Sloof, 1983; Sloof & Canton, 1983; Sloof et al, 1983*a*).

The spread of industry into areas of unbroken forest can produce the same complicating aspect, especially where spray operations against forest pests are already being carried out. For example, in New Brunswick in eastern Canada, Atlantic salmon and brook trout form the basis of commercial and recreational fisheries. From the early 1950s, salmon and trout streams in forested areas have been affected by aerial spraying against forest pests, firstly by DDT from 1952 onwards, then by its main replacement, fenitrothion since 1968. Within that period, other changes have taken place as a consequence of forest-based industry such as pulp and paper mills, and by increased use of fertilisers and herbicides in intensified forestry practice (Elson, Saunders & Zitko, 1972). To this must be added heavy copper–zinc contamination of waters since 1960 from the mining industry, The impact of pesticides is not only liable to be augmented or masked by the presence of these other contaminants, but may also be greatly influenced by such indirect effects of all these activities as faster run-off and erosion, and silting of stream beds, all a consequence of reduced forest cover. In contrast to the very great deal of research which has been directed to the impact of combined toxicants on fish, as illustrated by the guidelines issued by the European Economic Commission (Alabaster & Lloyd, 1980; EEC, 1981, 1984), the reaction of aquatic macroinvertebrates to such situations is only now at a late stage receiving the attention it merits.

II

THE ROLE OF LABORATORY
AND EXPERIMENTAL METHODS
IN EVALUATION

THREE

LABORATORY EVALUATION TECHNIQUES

INTRODUCTION

In comparing the reactions of different aquatic organisms to toxic chemicals (whether they are inorganic in the form of heavy metals, ammonia, cyanides etc. or more complex organic pesticides), it has long been the aim of the toxicologists involved to attain some standardisation or uniformity in the experimental techniques they devise. This approach to unification of methods has been particularly marked in the case of freshwater fish, for which group there are now standard methods acceptable in most countries.

These standard methods lay down guidelines on the exact conditions of the tests *vis-à-vis* water quality, temperature, number and size of test organisms, duration of exposure to the chemical dilution and, finally, the criteria to be adopted when measuring the effect in terms of immobility or mortality.

The need for the same degree of uniformity in testing other forms of freshwater life, invertebrates in particular, though slow to develop has gained increasing momentum in the last 10 years or so. Much of this incentive has come from the US Environmental Protection Agency (EPA) which now plays an international role in the screening and clearance of all pesticides and other toxic chemicals likely to have a harmful effect on the environment. One of the first fruits of this increased interest in aquatic invertebrates appeared in 1972 in the form of an exhaustive summary – compiled by the EPA – of all published information on their reactions in the laboratory to a wide range of pesticides, mainly insecticides, and herbicides (NTIS, 1972). This was followed in 1975 by *Methods for Acute Toxicity Tests with Fish, Macro-invertebrates and Amphibians*, (NTIS, 1975). These guidelines recommend that 'experimentally, 50% effect is the most reproducible measure of the toxicity of a toxic agent to a group of test organisms, and 96 hours is often a convenient, reasonably useful exposure duration'. This is expressed in the form of the 96-hour median lethal concentration (96-h LC_{50}), i.e. the concentration of toxicant in dilution water that is lethal to exactly 50% of the test organisms during continuous exposure for a specified period of time. The concentration of active ingredient (AI) of the toxic chemical is normally

23

expressed in parts per billion (or ppb, or its equivalent, micrograms per litre $\mu g/l$) or, sometimes, in parts per million (ppm, or its equivalent, mg/l).

Because of its ease of culture and handling in the laboratory, the water flea, *Daphnia*, has had a long-established role as a representative invertebrate in such laboratory tests. In this case it was recommended that a 48-h exposure was more acceptable, and that the most useful measure of effect was not mortality – sometimes difficult to assess exactly – but immobilisation. This is expressed as the 48-h median effect concentration, or EC_{50}. These criteria are based on the 'concentration' of toxic agent, i.e. the concentration in the test solution, rather than on 'dosage' which is something quite different, namely the amount of toxicant that enters the test organism.

In the exhaustive summary published by the NTIS (1972), the aquatic macroinvertebrates most commonly tested, in addition to *Daphnia*, and recorded in the US were the amphipod crustacean *Gammarus* ('scuds'), the freshwater shrimp *Hyalella* and the isopod *Asellus*; among insect aquatic stages are the stonefly nymphs of *Pteronarcys* and *Acroneuria*.

The EPA guidelines then proceed to distinguish the four main techniques for carrying out acute toxicity tests which are applicable to aquatic macro-invertebrates.

1 In the static technique test solutions and test organisms are placed in test chambers and kept there for the duration of the test.
2 The recirculation technique is like the static technique except that each test solution is continuously circulated through an apparatus to maintain water quality by such means as filtration, aeration and sterilisation and then returned to the test chamber.
3 The renewal technique is like the static technique except that the test organisms are periodically exposed to fresh test solution of the same composition, usually once every 24 hours, either by transferring the test organisms from one test chamber to another or by replacing the test solution.
4 In the flow-through technique, test solutions flow into and out of the test chambers on a once-through basis for the duration of the test. Two procedures can be used. In the first, large volumes of the test solutions are prepared before the beginning of the test and these flow through the test chamber. In the second (and more common) procedure, fresh test solutions are prepared continuously or every few minutes in a toxicant-delivery system.

With any of these techniques a pump or stirrer can be used to create a current in the test chambers to accommodate particular test organisms, but this will often increase aeration and volatilisation.

These four groups of techniques will form the basis of the present review. Without losing sight of the central theme of this book, namely the macro-invertebrates of running waters, consideration must also be given at this stage to information about certain test organisms whose habitats are the still waters of ponds, marshes and lakes. This is because it was studies on those organisms which provided the bulk of the early information about test methods and test criteria which have subsequently had a strong influence on the later work on stream invertebrates themselves.

The whole rationale of toxicity testing for freshwater organisms has been exhaustively examined more recently by Canadian authorities on fisheries and aquatic science (Scherer, 1979), providing a valuable manual complementary to those of the US Environmental Protection Agency. The Canadian authorities have also followed this up by producing a manual for the culture of selected freshwater invertebrates (Lawrence, 1981), a timely reminder that progress in knowledge of laboratory culture methods, and about the specific requirements of the various indicator organisms, will continue to play a vital role in strengthening the laboratory phase of evaluation.

STATIC TECHNIQUES

The use of the static technique is perhaps best illustrated by studies carried out at the Shell Research Laboratory, Sittingbourne, Kent on a range of aquatic invertebrates, including several forms characteristic of running water such as the crustaceans *Gammarus pulex* and *Asellus aquaticus*, and nymphs of the mayfly *Cloeon dipterum* (Stephenson, 1982). The chemical being studied was the synthetic pyrethroid cypermethrin, used widely in controlling a variety of insect pests. The exposure period in these tests was 24 h. Ten individuals of each invertebrate species were exposed to each of a range of concentrations, particular regard being paid to the stage of the life cycle and the approximate size of individuals tested. Two toxic effects were recorded, namely reduced motility after 2 h and 24 h, and death after 24 h, these recordings being used to calculate the EC_{50} and the LC_{50} values respectively.

The concentration of cypermethrin in duplicate test vessels containing 200 ml of the insecticide dilution was chemically analysed at the beginning and at the end of the 24 h exposure period. The tests were carried out in temperature controlled rooms, and the oxygen content of the water in controls and at the highest treatment level was measured at the beginning and at the end of the 24-h period.

The chemical assays carried out provided concrete proof of one of the potential variables in such static tests, namely that there can be considerable loss of chemical in the course of the test period. In this instance, losses of up to 30 % occurred when test concentrations were > 1 µg/l, while, at even lower test concentrations of 0.05 µg/l, losses as high as 70–80% were recorded. Loss of toxicant from static water tests is a well-known phenomenon accountable for in many cases mainly by adsorption on surfaces. That, together with the extremely high dilution of cypermethrin used in these tests may well have intensified this effect.

The attraction of static tests is that they are comparatively simple to carry out and to replicate under apparently strictly controlled experimental test conditions, of temperature, water composition, hardness, pH etc. In addition to the liability to loss of toxicant during the exposure period – a loss, incidentally, which must be accentuated when longer exposure periods of 48 and 96 hours are used – there is, however, another possible source of error in reaching the LC_{50} value. This is the important factor of variable control mortality, particularly where test organisms – not available in laboratory culture – have to be collected from the natural aquatic habitats and brought back in a live and healthy condition to the laboratory. The processes of collecting, handling and pre-test maintenance may all combine to weaken the test organism.

In the cypermethrin tests mortality and loss of mortality in controls did not exceed 10 %, but in other test series elsewhere control mortalities have on occasion been high enough to cast serious doubt on the validity of the test data. This is well illustrated by static tests carried out on stream macroin-vertebrates as part of the fenitrothion evaluation programme (see Chapter 7; Wildish & Phillips, 1972). The exclusively stream fauna used for those tests were collected on a drift screen held downstream while the substrate was disturbed. Exposures were either continuous, with acute lethal figures – LC_{50} being recorded after 24, 48 and 96 h, or discontinuous, involving shorter exposures of 1–10 h followed by a holding period of 86 h in clean water. In the continuous tests, mortality, particularly of dragonfly nymphs (Odonata) and stone fly nymphs (Plecoptera) in controls indicated that the static bioassay conditions imposed an additional stress on test organisms. In the case of the plecopteran (*Acroneuria*), the LC_{50} beyond 24 h could not be calculated because of control mortality – up to 60–80% in some cases – by the end of 96 h.

The discontinuous static tests were designed to simulate more closely the kind of exposures experienced in the field; short bursts of high concentration produced by the aerial spraying operations, being followed by rapid dilution due to the inflow of clean water. Even in this series, frequent control mortalities of one *Acroneuria* in five indicated an unacceptably high control mortality of 20%.

Some difficulties in the interpretation of static test data were also experienced in studies on the effect of the selective piscicide TFM and the piscicide/molluscicide Bayluscide (niclosamide) on non-target invertebrates (Kawatski, Dawson & Reuvers, 1974; Kawatski, Ledvina & Hansen, 1975). The particular test species in this series were fourth stage midge larvae, *Chironomus*. The criteria used to judge toxic effects was the EC_{50}, immobility of test organisms being noted after 8, 24, 48, 72 and 96 h in a continuous

26

exposure test. In the course of these tests, it was noted that some chironomids immobilised by TFM and apparently dead, revived when they were placed in toxicant-free water. Post-exposure observations were able to establish that the concentration necessary to produce death was much greater than that which produced immobilisation – immobilisation which had hitherto been assumed to by symptomatic of death – the 8-h LC_{50} being 12 times greater than the immobility-based EC_{50}. These workers concluded that, in the light of these observations, and the difficulty in defining death, particularly in small and/or normally sluggish invertebrates, the LC_{50} value in short-term exposures can be accurately determined only if the affected organisms are held in toxicant-free water after the exposure period.

Somewhat similar conclusions were reached in another study on the toxicity of TFM to select aquatic invertebrates (Smith, 1966). Mortality in these continuous exposure static tests was normally determined at the end of the 22–24 h test period. In the case of leeches, snails and clams however the effect of TFM was to make them sluggish during testing to such an extent that mortality could not be determined with certainty at the end of the test period. These organisms therefore were allowed a recovery period of 8–10 days in fresh water before any decision about mortality was made.

STATIC TESTS ON STREAM MACROINVERTEBRATES

One of the earlier applications of the static test to exclusively stream invertebrates showed how this technique can reveal distinct differences in reaction between closely related genera (Jensen & Gaufin, 1964). These were studies carried out near Salt Lake City in Utah following the reports in 1955 of the disastrous effect on aquatic life caused by DDT spraying against spruce bedworm in large areas of forest in the Yellowstone National Park and other parts of the Rocky Mountain region. The laboratory toxicity tests concentrated on 2 species of stonefly nymph, *Pteronarcys californica* and *Acroneuria pacifica*, selected because of their importance as food for trout and other sport fishes in the Rocky Mountain region, and as indicators of clean water. In addition, they were easy to obtain at all seasons of the year and to maintain in the laboratory. The reactions of these two species were tested to a wide range of chlorinated hydrocarbon insecticides, DDT and dieldrin etc., and organophosphorus compounds such as parathion and malathion.

The static test method consisted simply of preparing various dilutions of the chemical in water taken from a natural creek, adding the stonefly nymphs and then making observations at selected times on their reactions throughout the following 4 days (96 h). In addition to observations on the mortality rate during that period, the sequence of symptoms displayed by the animals before

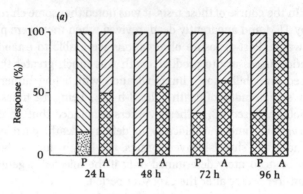

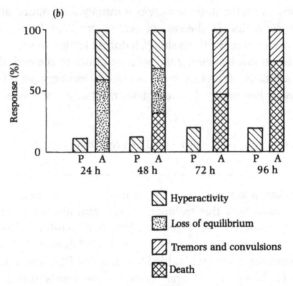

Hyperactivity

Loss of equilibrium

Tremors and convulsions

Death

Fig. 3.1. Example of early use of static toxicity tests to compare the sensitivity of nymphs of the stoneflies *Pteronarcys californica* (P) and *Acroneuria pacifica* (A) to two organophosphorus insecticides (a) parathion and (b) malathion, based on progressively increased exposures to a fixed concentration – 0.018 ppm (after Jensen & Gaufin, 1964).

death was recorded using five symbols. For both chlorinated hydrocarbons and organophosphates, *A. pacifica* was much more sensitive than *P. californica*, and the two species also showed a different reaction pattern to each insecticide with regard to onset of hyperactivity, loss of equilibrium, appearance of tremors and convulsion ending in death (Fig. 3.1). With regard to the progressive nature of these symptoms, it was noted that those stonefly nymphs which had reached the stage of showing tremors and convulsions failed to recover if removed from the insecticide dilution and transferred to clean water. If they were removed prior to the onset of tremors, a return to normal behaviour occurred in more than 90% of the specimens.

The mortality data in that series were expressed in the form of TL_m (median tolerance limit), which is another way of describing the LC_{50}. Static tests also played an important part in showing how one and the same test organism could reveal an enormously wide range of susceptibility to different insecticides (Sanders & Cope, 1968). This is well exemplified by nymphs of the stonefly *Pteronarcys californica* in which the 24-h LC_{50}, in parts per billion (ppb) for a range of chlorinated hydrocarbons was as follows:

endrin	4
dieldrin	6
endosulphan	24
DDT	41
chlordane	170
TDE(DDD)	3000

and, for some organophosphorus insecticides,

parathion	28
Dursban	50
Abate	120
fenthion	130

STATIC TESTS EMPLOYING RECYCLING

Recycling techniques, as distinct from through-flow methods described below, have not been used extensively in pesticide-impact studies on aquatic invertebrates. However, they have been rather more widely used in culture methods, particularly to meet the demanding requirements of such sensitive organisms as *Simulium* larvae (Raybould & Grunewald, 1975; Brenner & Cupp, 1980) and some of these recycling principles have been incorporated in laboratory experimental tests with pesticides, described later.

A good example of this principle applied to another test animal with special requirements is provided by studies on the impact of the larval lampricide, TFM, on nymphs of the mayfly *Hexagenia* (Fremling, 1975). In the US, *Hexagenia* nymphs play a vital ecological role in lakes, rivers and streams. The nymphs are important to fish because they convert organic detritus, algae and bacteria into high-quality fish food. Since *Hexagenia* nymphs prefer silt bottoms where they can construct burrows, they usually inhabit the same type of habitat, such as silty streams, used by larval lampreys (*Petromyzon marinus*) which are the object of control operations with TFM.

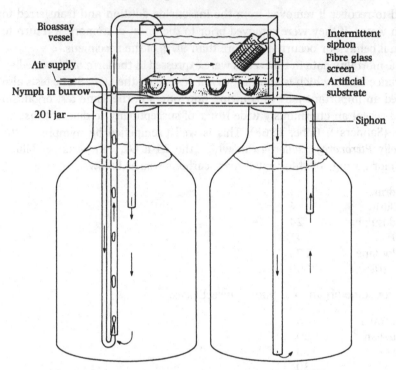

Fig. 3.2. Recycling bioassay apparatus with an artificial burrow substrate for use with *Hexagenia* mayfly nymphs, as used in tests with the larval lampricide TFM (after Fremling, 1975).

Hexagenia are good test organisms because they are easily collected and cultured. As these nymphs tend to abandon their burrows when under stress, it was necessary to devise a recycling test vessel which allowed for the special requirements by providing the necessary environmental conditions. This was achieved by means of the equipment shown in Fig. 3.2. Each test vessel took the form of a rectangular glass aquarium provided with a substrate 23 cm long, 5 cm wide and 5 cm deep. The substrate contained 10 U-shaped burrows, providing semi-darkness and seclusion, and suitable for each nymph to settle into. Constant recirculation of test water was achieved, not by the conventional pump, but by an ingenious airlift arrangement provided by the steady incoming air supply. This pumped the water from a 20 l glass jar to another similar one, via the test vessel. An arrangement of syphons ensured automatic recirculation of the water. The necessary amount of the toxicant was added to each of the 20 l glass jars giving a combined volume of water of 38 l, and allowed to recirculate completely before the experiment started.

LC_{50} values were calculated after 6, 12, 24 and 96 h of continuous exposure tests; these were carried out in both hard water and soft water, and at two controlled temperature ranges, 17–18 °C and 24.5–26.5 °C. The satisfactory nature of this special technique was evident in that of 10 96-hour tests, control mortalities were kept down to 10% in 5 and to 0 in 2. In all tests there was a marked tendency for treated nymphs to abandon their burrows for varying lengths of time before they finally succumbed, a behaviour pattern not unlike that of the case-bearing caddis larva, *Brachycentrus*, in experiments involving other chemicals (Chapter 5).

CONTINUOUS THROUGH-FLOW TESTS

The many variables and limitations of the static test, particularly when extended over a period of 4 days, have long been recognised in the general field of aquatic toxicology. Possible loss of chemical due to volatilisation, adsorption or degradation may be accelerated if the test organisms require aeration – as they often do – in order to survive. Under static conditions too there may be a harmful accumulation of excretory products. It is only comparatively rarely, as in the example cited on page 25, that possible fluctuations in concentration of the toxic chemical during the course of the test have actually been checked directly by chemical analysis.

Awareness of these various uncertainties early underlined the need for continuous flow tests in which the concentration of toxicant could be maintained at a constant level throughout the complete test. In order to achieve this, much more sophisticated apparatus had to be designed to ensure that mixing of pesticide, its serial dilution and the final delivery of precise concentrations to a series of test vessels could be carried out on the large scale demanded by many laboratories. Much technical ingenuity has gone into designing the serial dilution equipment and flow-through apparatus which can ensure the controlled flow rate desired (Mount & Warner, 1965; Mount & Brungs, 1967; Burke & Ferguson, 1968; Brungs & Mount, 1970; McAllister, Mauck & Mayer 1972; Chandler, Sanders & Walsh, 1974; Smith & Hargreaves, 1983).

The ingenuity of different workers and their different approaches to the same basic problems have expressed themselves in the wide range of equipment and experimental procedure. The following examples selected are mainly from the field of fish toxicology but, in view of the fact that experience gained in that older discipline has had a powerful influence on the much younger science of invertebrate toxicology, there are useful lessons to be learned. With regard to the actual flow rate produced by different types of equipment, some of these were designed to discharge known concentrations of toxicant into each test vessel at rates up to 400 ml/min (Mount & Brungs,

1967). In another case, in which 10 l test vessels were used, a flow rate of 300 ml/min was established (Burke & Ferguson, 1968). In still another case, which was long-accepted standard practice in the UK (Alabaster & Abram, 1965) and employed *Rasbora* as the standard test fish, 100 ml of freshly prepared test solution passed through the 500 ml test chamber every 10 min, i.e. replacing the test solution at the rate of 10 ml/min. There have also been considerable variations in the size of the test vessels or aquaria used, as exemplifed by the following:

Size of test vessel	Reference
15 l tanks } 37 l tanks }	(Arthur *et al.*, 1974)
40 l aquaria	(Abram, 1973)
10 l vessels	(Burke & Ferguson, 1968)
500 ml vessels	(Alabaster & Abram, 1965)

The different discharge rates used according to different designs of equipment are not so important in themselves, but in relation to the size of the test vessel into which the pesticide dilution is delivered. The important point is the time taken by the delivery system to replace completely the water in the test vessel. Here again, the EPA (page 23), has provided guidance in stating that the toxicant delivery system should provide at least 10 water volumes in 24 h, and that the flow rates through each test vessel should not vary by more than 10% within tests and between tests. With adherence to this general directive, the actual size of the test vessel should not be critical and can be as small or as large as befits the requirements of the particular type of test organisms.

In well-established fish toxicology laboratories providing ample resources and expertise, these test methods have on occasion been successfully translated directly to the special problems of aquatic invertebrates (Sanders & Walsh, 1975). In general, however, it is doubtful if such elaborate and staff-demanding equipment for ensuring continuous through-flow has a really significant role to play in purely invertebrate studies. From the steady but slow rate of flow, and slow replacement of toxic dilution in the test chambers quoted above, it cannot be claimed that the test organisms are being exposed to anything approaching the degree of water flow they are accustomed to in their natural running water habitat. This need to re-create, under experimental conditions, the type of rapid flow and water change best suited to stream invertebrates had led to the development of techniques specially designed for macroinvertebrates along new independent lines from those practised by fish toxicologists.

This need has found its most useful expression in recent years along two

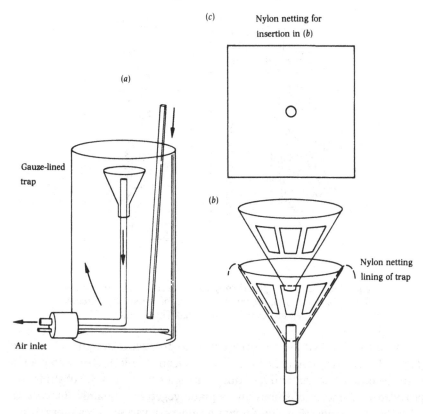

Fig. 3.3. (*a*) Rapid through-flow test vessel (adapted from 10 l
aspirator bottle with top cut off) showing inflow/outflow
arrangement and particulars of (*b*) fenestrated funnel and (*c*)
netting exit trap designed to minimise pressure of outflow on
small invertebrates (after Muirhead-Thomson, 1978*c*).

different lines, namely the development of rapid through-flow tests specially
designed for a range of stream macroinvertebrates, and the miniaturisation
of laboratory-stimulated streams or experimental channels.

THE RAPID THROUGH-FLOW TECHNIQUE

The rapid through-flow technique is best illustrated by methods which have
been found equally applicable to a variety of stream organisms associated
with both African tropical waters (Zimbabwe) and with European running
waters (Muirhead-Thomson, 1973, 1978*b*,*c*, 1981*b*).

The basic test chambers in this technique are 10 l glass aspirator bottles

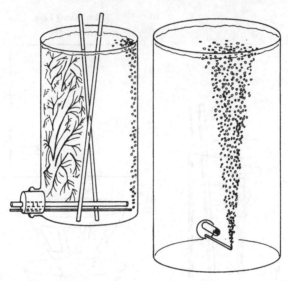

Fig. 3.4. Use of rapid through-flow test vessel to include *Simulium* larvae, aggregated on wall of vessel by directed compressed-air jet (after Muirhead-Thomson, 1977).

in which the rounded top has been cut off to produce a straight-sided vessel (Fig. 3.3). The arrangement for inflow of clean dechlorinated water, or of pesticide dilution water, and for outflow are shown in Fig 3.3. Substrates are provided in the form of stones and aquatic vegetation (mainly *Ranunculus*) taken from the collecting site. Known numbers of various stream organisms can be introduced either as a pure sample of single species or of several species together. In order to include *Simulium* larvae (with their special requirements) in this kind of test, vegetation with attached larvae collected from the natural habitat is placed in static water in the test vessel prior to introduction of any other test organisms (Fig. 3.4). By means of a jet of compressed air directed against the side of the vessel, larvae are induced in the course of the following few hours to leave their attachment on the vegetation and migrate to the vertical zone of maximum aeration and agitation. On the following day, when several hundred larvae may have become firmly fixed and aggregated on the wall of the vessel, the vegetation is removed along with other debris. Other test macroinvertebrates can then be added as required, with appropriate substrates, and a rapid through-flow of clean water is introduced.

This technique was originally developed to deal with the specific problem of pesticide impact following short, aerially applied, field applications of 10–15 min at an application point on rivers or streams. The standard adopted for routine through-flow tests was 1 h exposure followed by 24 h in clean flowing water. During the course of that hour, manually prepared serial dilutions of the pesticide (*Simulium* larvicides in this particular series) freshly prepared in a series of 20 l glass aspirator bottles, flowed through each test vessel at the rate of 60 l/h, i.e. approximately 1 l/min. With the introduction of pesticide dilution water, this flow rate was increased long enough to reduce to about 5 min the initial lag before the clean water in the vessel is completely replaced by the pesticide dilution. Because of this initial lag associated with any replacement technique (Sprague, 1969) this test was not used for studying the effects of exposures of less than 1 h in duration. For accuracy in studying exposures in the 5–15 min range, the more precise simulated-stream technique is essential, as is described below.

In this 10 l test vessel, different organisms can be introduced, with the appropriate substrates, to form a miniature ecosystem in which community reactions to introduced toxic chemicals can be studied within a single test. Such invertebrates as *Gammarus* (Amphipod), *Hydropsyche* and *Brachycentrus* (Trichoptera) *Baetis* (Ephemeroptera) can coexist successfully along with the established *Simulium* larvae, and can all be tested together. This type of preliminary test usually provides the first indication of any marked specific differences in reaction, differences which can subsequently be established more precisely by exposing single species samples to the appropriate range of pesticide concentrations. Some results obtained by this procedure are shown in Table 5.2. These reveal for example the high sensitivity of *Baetis* spp. to most larvicides tested, and the wide range of tolerance levels shown by other species, reaching an extreme example in the case of the high tolerance shown by *Gammarus* to the organophosphorus insecticide Abate.

This manually operated test was geared to the capability of a single operator who made up the necessary concentration of insecticide dilution in the three 20 l aspirator bottles necessary to provide the 60 l flowing through the test vessel in the course of 1 h. For longer exposures of 24 hours and upwards, for which this equipment is well designed, more sophisticated automatic dosing would clearly be desirable (see page 31). Despite the continuous rapid through flow of water, or pesticide dilution, the limitations to this and other closed-system types of test vessel are that they cannot clearly define such behavioural reactions as increased activity or detachment which could lead to downstream drift. In order to study this aspect, it is essential to create a laboratory stream or experimental channel on the lines described below.

EXPERIMENTAL CHANNELS AND SIMULATED

STREAMS

GENERAL FRESHWATER BIOLOGY

It has long been recognised by freshwater biologists that certain aspects of growth, metabolism and community structure can only be analysed experimentally in model streams or artificial channels. It has increasingly been recognised too that culture methods for stream macroinvertebrates demand a more realistic approach to the simulation, in the laboratory, of the natural flowing water environment to which these organisms are adapted. An upsurge of interest in these needs has led in the last 25 years to the development of a variety of different model streams based on the 'closed' system in which recirculation is produced by an electric pump rather than an 'open' system in which an outside source of clean water flows through the channel on a once-through basis (Lauff & Cummins, 1964; Whitford, Dillard & Schumacher, 1964; Craig, 1966; Feldmeth, 1970; Fikes & Tubb, 1972).

The general principle of these is well exemplified in Fig. 3.5 (Craig, 1966), which incorporates ideas from different workers. The fibreglass centrifugal pump, powered by an electric motor, is capable of directing 3180 l of water per hour into a straight perspex trough 91.5 cm long by 15 cm wide and 15 cm deep. The trough floor is provided with small ridges to retain substrate in the form of gravel or small stones, and has a nylon gauze barrier at the outlet to retain organisms in the trough. From the channel, the water flows through a filter of activated carbon and then into a reservoir in which heating and refrigerator units control temperature. Controllable fluorescent-tube lighting allows growth of algae and other flora. This technique has been used successfully to rear a wide range of aquatic invertebrates.

Increased interest on the part of freshwater biologists in creating artificial streams in which the many variable factors in the natural habitat can be controlled and tested experimentally, has expressed itself in its most impressive form within the last few years. This is the large recirculating channel developed by scientists of the Freshwater Biological Association's (FBA) River Laboratory at East Stoke, England, and is the first attempt to design and build a more or less complete ecosystem in a channel with a recirculating facility (Ladle, Welton & Bass, 1980; Ladle *et al.*, 1977, 1981; Marker & Casey, 1983; Pinder, 1985). The channel is 'race-track shaped', 53 m in length with a basal width of 1 m and a top width of 2 m (Fig. 3.6*a*). Water is supplied from a chalk aquifer, and is recirculated by a screw pump. The principle involved is the re-creation of a new natural stream. The channel is filled with flint

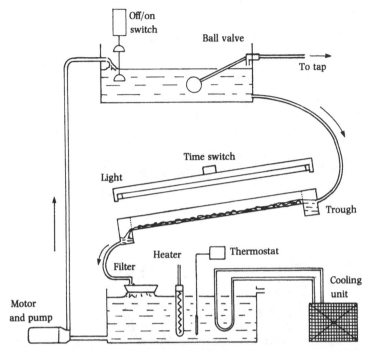

Fig. 3.5. Design of an artificial experimental stream using closed circuit recirculation system, suitable for wide range of stream invertebrates (after Craig, 1966).

gravel from a terrestrial source, but of similar character to that found in small chalk streams. Six stones from an adjacent chalk stream provided the inoculum of algae. Water velocity could be maintained at 0.4 m/s.

The main objective of this device was to study the development of a stream invertebrate community from scratch, increases or changes in the population being followed by means of a total of 620 cylindrical invertebrate samplers, filled with introduced gravel, and buried level with the surface of the substrate (Fig. 3.6b). Drift occurs in this channel but, as it is a closed system, there is no gain of invertebrates from either 'upstream' or 'downstream'. The recirculating stream proved to be ideal for several species of Ephemeroptera, including *Baetis* and *Ephemerella* (Welton, Ladle & Bass, 1982), as well as net-spinning caddis larvae, *Polycentropus* (Bass, Ladle & Welton, 1982).

Although large through-flow troughs have been used in Zurich for a number of years in the investigation of self-purification following pollution (Wuhrmann, 1964; Wuhrmann, Ruchti & Eichelberger, 1966), nothing on the scale of the FBA model has yet been attempted in studies on pesticide contamination. The results of that experimental work however are far from purely academic. Quite apart from the lessons to be learned from all the

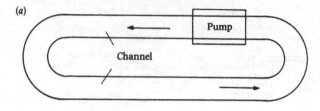

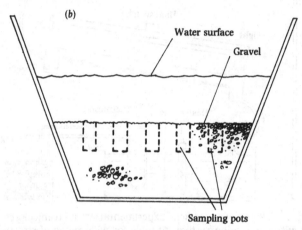

Fig. 3.6. (*a*) Large recirculating channel as used by the UK Freshwater Biological Association (after Ladle *et al.*, 1980). (*b*) Cross section of recirculating channel showing invertebrate samplers buried in substratum. (Ladle *et al.*, 1980).

technical features involved, valuable information has been provided about the build up of invertebrate populations from zero, information which has relevance to the recolonisation of stream fauna decimated by the passage of insecticide.

EXPERIMENTAL CHANNELS AND SIMULATED STREAMS IN *SIMULIUM* ECOLOGY

In the field of pesticide impact on stream macroinvertebrates, the development of experimental channels has been particularly marked in the case of blackfly (*Simulium*) larvae. Adult blackflies are important pests whose biting habits in some countries create an intolerable nuisance for humans and for domestic stock. Even more important is their role as vectors of a human parasitic disease, onchocerciasis, in tropical Africa and certain countries of Central America. In areas of high transmission, such as West Africa, the heavy parasite load in infected humans may produce ocular complications leading to the conditions commonly known as 'river blindness'. Long experience has shown that control of these biting fly populations can best be achieved by measures directed against the larvae whose habitat is the swift-flowing waters of streams and rivers. In order to develop appropriate laboratory screening methods to test the effect of various chemical larvicides, special measures must be taken to meet the special requirements of *Simulium* larvae. These larvae are normally very firmly attached to substrates such as stones, trailing or submerged vegetation, etc. by anal suckers. When removed from running water, or immersed in still water, the larvae readily become detached and float away at the end of fine spun threads. These features create certain difficulties in handling by conventional methods in the laboratory. Larvae firmly attached to a substrate cannot be dislodged or removed by pipette without some risk of damage – some species being more refractory than others in this respect – and, once larvae have become detached, they tend to aggregate and form communal webs which, again, makes handling difficult. Associated with all these special adaptations to the fast-flowing water of their natural habitat are the many obstacles to be overcome in transferring live larvae from field collection sites to base laboratory, and in maintaining them in healthy conditions once they get there. As an essential requirement of all critical laboratory tests is to maintain control mortalities at an acceptably low level, these difficulties have to be overcome before any serious evaluation programme can be contemplated.

The search for the ideal laboratory evaluation technique has developed along many different lines corresponding to the many different workers involved. These divergent techniques have evolved more thanks to the ingenuity of each investigator, and the laboratory space and amenities available, than to any great variety in the requirements of different *Simulium* species. Indeed, throughout their wide geographical distribution, the great majority of *Simulium* larvae which live attached to natural substrates – in distinction from those which are specially adapted to using living aquatic animals to attach to – react in essentially the same way in captivity. Techniques for example originally devised for West African species were later found to be applicable without modification to European species, and even to an entirely different fauna in southern Africa (Muirhead-Thomson, 1973).

One of the first running water evaluation techniques to be developed for testing *Simulium* larvicides was simply to introduce substrates with attached larvae from a natural stream into a trough with running water, and allow as many larvae as possible to re-attach in the artificial stream (Travis & Wilton, 1965; Wilton & Travis, 1965; Travis & Schuchman, 1968). This idea took two distinct forms; in the first, the portable troughs were set up close to the natural habitat and supplied with natural stream water; in the second, larvae were transferred from the field to troughs installed in the laboratory.

A good example of the first variation was a series of tests carried out with a North American species, *Simulium pictipes*, which frequently occurs in conspicuous dense masses on flat rock covered by a thin sheet of fast-moving water. In this case it was found that larvae could be dislodged gently and collected without suffering damage in a sieve as they were swept downstream. Larvae were then introduced into a series of V-shaped metal troughs 1.83 m long, lined first with polyethylene sheeting, with brown wrapping paper superimposed. The paper liner provided a good, disposable, surface for *Simulium* larvae to attach. When sufficient numbers of larvae – normally 20 in each trough – had attached, well-mixed chemical dilutions were monitored in from reservoir tanks through the channels. Dead or detached larvae were collected in a sieve arranged in such a way that larvae were retained in a pool of flowing water. After exposure to larvicide flow for a standard period, larvae surviving in the troughs were removed by means of a camel hair brush and placed in the nylon-lined sieve. The nylon was then formed into a close bag which was immersed in the water of the stream for 24–48 h observation. A 15-min exposure period was adopted as standard, and 20 mature larvae of equal size used in each test.

The need for testing the effect of chemicals on *Simulium* larvae under flowing water conditions led to the development of another type of trough technique (Frempong-Boadu, 1966; Jamnback & Frempong-Boadu, 1966), simple enough to have stood the test of time, and still used in the laboratory phase of evaluation in the Onchocerciasis Control Programme (OCP) based on *Simulium* control by larvicides over entire river systems.

These troughs are 91 cm long, 30 cm wide and 15 cm deep, with an effluent lip 15 cm wide and 10 cm long. Stones heavily populated with larvae collected from a natural habitat were immersed in the trough which was supplied from a spring-fed pond. In this process many larvae became detached and were carried downstream out of the system; but a sufficient number became aggregated at the zone of swiftly flowing water at the lip of the trough, where they could be observed easily through the thin film of smooth-moving water, less than 1.25 cm deep. The technique had the great advantage in that no handling of larvae was involved, ensuring that no injury or trauma was sustained. Also, numbers of larvae, and stages, could be standardised for each test simply by removing surplus larvae from the lip, to leave the normal 25–35 medium-sized larvae required. Larvae which became detached as a result of exposure to insecticide introduced into the trough were trapped in cloth sieves and transferred to clean aerated water for a 24-h holding period. The standard exposure period initially adopted was 5 min.

With 24 of these troughs installed in a spacious fish hatchery there was ample scope for replication and for the testing of several different chemicals and formulations at the same time.

These trough tests demonstrated the great advantage of using running water techniques for studying the reactions of stream fauna like *Simulium* larvae which are adapted to life in fast-flowing waters. Only in the trough or experimental channel is it possible to assess the important reaction of detachment, as distinct from mortality, and only in the rapid flow of water do larvae continue their normal feeding movements of the mouth brushes.

STUDIES ON MACROINVERTEBRATES IN GENERAL

In the course of studying the problem of establishing *Simulium* larvae in a controllable laboratory situation, a technique devised by the author has undergone progressive changes which provide an interesting demonstration of the evolution of ideas and their modification to meet a much wider range of requirements than originally visualised.

The prototype test, originally designed for the larvae of *Simulium damnosum* and other West African species used a simple 10 l glass battery jar. The vessel was filled with water and a jet of compressed air was directed against the inner wall of the jar, producing a vertical zone of maximum aeration and agitation (Muirhead-Thomson, 1957). Aquatic or trailing vegetation with attached larvae collected from a natural stream was immersed in the water in the vessel, and larvae allowed to move spontaneously from the vegetation to the wall of the vessel over a period of several hours. Eventually the larvae formed a dense vertical band, containing up to a thousand larvae on occasions, along the line of the rising air bubbles (Fig. 3.4). With the removal of vegetation, the vessel in this simple form served a useful purpose in an exploratory series of observations on the reactions of larvae to different concentrations of insecticide introduced.

The next stage of development in technique was to expose the aggregation of larvae – firmly fixed to the wall of the vessel – to a current of water and not just to water movement induced by the jet of air bubbles (Muirhead-Thomson, 1969, 1973). This was achieved by rotating the vessel through 90 °C while at the same time introducing a flow of water into the lower inner end of the vessel (Fig. 3.7a,b). When this operation was carried out smoothly within about half a minute, the majority of larvae remained firmly attached, and were now exposed to a continuous rapid flow of water in a miniature simulated stream 25 cm × 4 cm × 5 mm deep. The process of decanting the 10 l of water during the rotation from vertical to horizontal was facilitated by either of two methods viz., temporarily attaching an extended lip to the top of the vessel, (Fig. 3.7c) or by siphoning off the water while the vessel was being decanted (Muirhead-Thomson, 1977).

When rotation to the horizontal position was complete, larvae left stranded on the wall of the vessel outside the narrow stream were induced back into the mini-stream by means of jets of water from a plastic wash bottle.

41

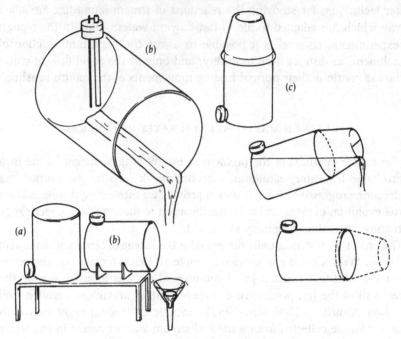

Fig. 3.7. Figure showing conversion of (*a*) vertical test chamber with aggregated *Simulium larvae,* into (*b*) horizontal simulated stream. (*c*) method for achieving smooth rotation through 90° by means of temporary plastic funnel fitment (after Muirhead-Thomson, 1977).

The main achievement of this technique was to establish, without handling at any stage, a band of several hundred *Simulium* larvae aggregated in the miniature simulated stream in much the same way as on substrates in their natural habitats. Through this stream, precisely controlled concentrations of chemical could be monitored for periods of 5, 15, 30 min or more, as required. Larvae which became detached following exposure to insecticide were trapped in a pool of water and subject to a continuous flow (Fig. 3.8). Insecticide dilutions were freshly made up in a series of 20 l aspirator bottles, and allowed to flow by gravity through the channel for measured periods, usually at the rate of 80 l/h; i.e. 1.3 l/min.

The next phase of development was to use the same miniature simulated stream for other stream macroinvertebrates. In order to do this, certain obstacles had to be overcome. When very active invertebrates such as the mayfly nymphs *Baetis* and the freshwater shrimp *Gammarus* are introduced directly into the flowing water in this stream, they are almost immediately washed downstream into the trap. To obviate this, and allow them to become established, the vessel was tilted from the horizontal so that a pool of water

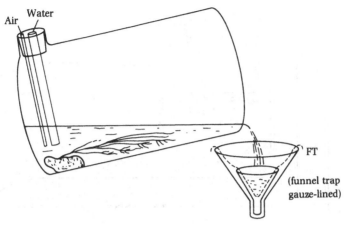

Fig. 3.8. Modification of miniature simulated stream to suit stream macroinvertebrates other than *Simulium* larvae. FT, details of water-retaining trap for drift organisms (after Muirhead-Thomson, 1978*b*).

was formed at the inner end (Fig. 3.8). The water flow was stopped, and a jet of air bubbles introduced into the pool. Known numbers of macroinvertebrates were then pipetted into the pool and provided with natural substrates in the form of small stones from the stream bed or fronds of aquatic vegetation. When the invertebrates had settled down in this environment, usually overnight, a flow of water was gradually introduced and the vessel gradually moved back to horizontal or near-horizontal position. The impact of different introduced chemical could now be studied in the same way as described above. Apart from its use in demonstrating different tolerance levels of different stream macroinvertebrates in a flowing water system, it was now also possible to demonstrate specific differences in behaviour reaction with regard to irritation and activation leading to detachment and downstream drift. The results obtained will be discussed in more detail in Chapter 5.

The small size and compact construction of the miniature simulated stream allows several systems to be set up in limited laboratory space. Each of the channels was well designed to test samples, in the range of 10 to 50 animals, of single species. The limited space and volume of water in the channel imposed restrictions on testing several indicator species together at the same time or to establish anything like a miniature stream ecosystem as distinct from the enclosed ecosystem set up in the rapid through-flow type of test vessel (page 33).

These needs led to the design of an improved and extended miniature simulated stream (Fig. 3.9; R. C. Muirhead-Thomson, unpublished results). A length of fibreglass guttering was secured to the upper part of the converted 10 l aspirator bottle which forms the basis of the original miniature stream.

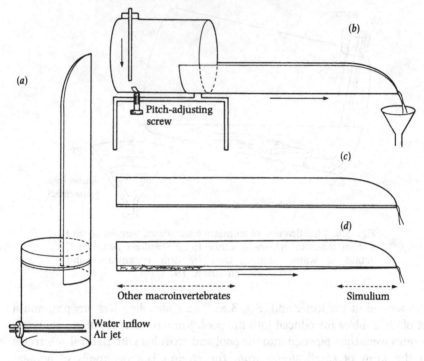

Fig. 3.9. Design of extended simulated stream, incorporating a variable-pitch principle (R. C. Muirhead-Thomson, unpublished results).

Simulium larvae were induced to leave their immersed vegetation substrate and attach to the wall of the vessel, as in the original design. When the apparatus was rotated through 90 °C, an extended channel of water, now 75 cm long, was formed (Fig. 3.9(b)). A screw attachment enabled the pitch of the channel to be altered smoothly and gradually from the horizontal so that the slope could be increased or decreased. A slight decrease in the slope had the effect of increasing water velocity in the outer half of the channel and led to an aggregation of *Simulium* larvae in the outer half or third of the channel (Fig. 3.9(d)). The slower flow and increased water volume in the inner half of the channel produced by the alteration of pitch now produced a *Simulium*-free zone in which there was ample space for other invertebrates to be introduced and attach to substrates provided, such as strands of *Ranunculus* rooted in gravel from the original stream bed.

The result of these improvements was that a wide range of stream macroinvertebrates could be established in the same miniature simulated stream in which *Simulium* larvae had already become attached, but in such a way that the part of the channel which they dominated did not overlap or

44

disturb the more rapid flowing *Simulium* zone near the lip of the miniature stream (Fig. 3.9 (*d*)).

A further improvement was that the supply of chlorine-free water continually flowing through the channel on a once-through basis no longer came from a large storage tank, but was directed first through a large fibreglass aquarium containing gravel, stones and rooted aquatic vegetation from the original stream. This innovation ensured, as far as possible, that the macro-invertebrates in the experimental channel were exposed to a continuous supply of well-aerated water containing natural food in the form of algae, bacteria and organic particles. In practice it was found that this supply aquarium remained functional for about 2 months in the naturally illuminated, glass-roofed insectary in which it was housed. After that time, increasing algal growth, which extended to the channels themselves, made it necessary to replace completely the contents of the supply aquarium.

The need for such a supply aquarium could of course be obviated if the laboratory is sufficiently near a natural uncontaminated stream to allow water from that source to be pumped, or flow by gravity, through the system. Such ideal conditions do occur occasionally (see below), but it is safe to assume that the majority of aquatic toxicology laboratories do not have this facility. A final advantage of extending the original miniature simulated stream in the form of an open channel was that it was now possible to set up a stereo microscope for close observation of attached macroinvertebrates *in situ*.

The testing procedure in this new variable-patch modification was precisely the same as in the original model, freshly prepared dilutions of pesticide at the desired concentration being monitored into the stream from a series of 20 l aspirator bottles, replacing completely the flow of clean water from the supply aquarium for the duration of the exposure period.

The advantages of a battery of small continuous-flow experimental channels are also patent in the system devised by researchers in New York State (Gaugler *et al*, 1980). In this system, each unit is designed to be used either as a closed system – water continually being recycled – or as an open system like the miniature simulated stream described above. In the former fish hatchery where these systems were installed, 36 such units were established, allowing considerable scope for increasing the range and replication of experiments. A diagram of the basic unit (Fig. 3.10) shows that it consists principally of a reservoir tub (*a*), a recirculating pump (*b*) with hose and valve (*c*), freshwater supply valve (*e*) delivery funnel (*g*) and a tray for larval attachment (*i, j*).

Whether used as an open or flow-through model, fresh water continually enters the delivery funnel from the supply valve, passes over the tray with attached *Simulium* larvae, and is allowed to overflow the reservoir tub, thence

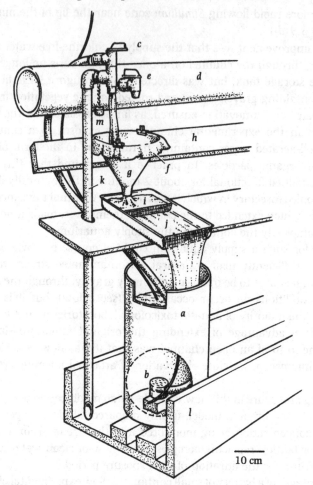

Fig. 3.10. Design of battery of small continuous-flow experimental channels (after Gaugler *et al.*, 1980). (*a*) reservoir tub, (*b*) recirculation pump, (*c*) recirculation value, (*d*) water supply pipe, (*e*) supply valve, (*f*) organdy cloth filter, (*g*) delivery funnel, (*h*) funnel support, (*i*) tray reservoir section, (*j*) tray attachment section, (*k*) standing waste pipe, (*l*) waste trough, (*m*) cap.

to a waste trough and a floor drain. In the recirculation or closed operation, the unit is filled with stream water from the supply valve. A pump then lifts the water from the reservoir tub and into the delivery funnel. The water flows down the funnel, over the tray, and is returned to the reservoir tub for recycling.

The tray component comprises two parts viz., the reservoir (*i*) and larval attachment (*j*) sections. Between these is a stainless steel screen whose

46

function is to convert the turbulent inflow into an even sheet of swiftly flowing water over the larval attachment area. This system also, as in the extended miniature simulated stream above (page 43) allows larval behaviour to be closely observed with the aid of the stereomicroscope.

In establishing *Simulium* larvae, the tray discharge end is propped up to allow a pool of water to collect at the rear half of the tray. Larvae from field collections are transferred according to numbers and size required to this pool by means of a brush. After this the flow rate is increased, and the tray gradually tilted back to normal – an operation essentially similar to that previously described for establishing non-target macroinvertebrates in the miniature simulated stream (page 42). For use with macroinvertebrates other than *Simulium* larvae, the tray is provided with appropriate substrates.

An advantage of this dual-purpose unit is that the recycling procedure can be adopted when constant water temperatures, above that of the inflow from the stream source, are required, simply by placing an aquarium heater on the reservoir tub.

Ideally, simulated-stream systems should enable stream macroinvertebrates to be tested at a range of constant temperatures; but in practice strict temperature control at figures above or below normal laboratory levels – or above or below inflow from a natural stream – would require rather elaborate means for maintaining constant temperatures in the large volume of water – from 80 to 100 l/h – flowing through the experimental channel on a once-through basis.

OBJECTIVES, POSSIBILITIES AND CONSTRAINTS OF

LABORATORY STREAM RESEARCH

As pointed out in the introduction to this chapter, the use of laboratory techniques – simulated streams in particular – for studying the reactions of aquatic macroinvertebrates is still in a very early exploratory stage. An examination of the various techniques developed by different investigators provides ample proof of the great diversity in approach, and of the fact that nearly all of these independently devised systems have certain advantages and certain shortcomings (Maciorowski & Clarke, 1980). However, the technical difficulties that have had to be overcome in the initial handling and maintenance of these sensitive stream organisms, to say nothing of the problem of devising valid toxicity techniques, can in no way detract from the firm conviction that such laboratory investigation is not only an essential component of the overall evaluation of pesticide impact, but can play an increasingly important role with improvement and refinement of the experimental approach.

Laboratory and experimental methods in evaluation

Perhaps this would be an appropriate place to take a stock taking of what can and cannot be achieved by controlled experiment, and to examine critically the application of laboratory data to the complex environment of the field habitat (Benfield & Buikema, 1980; Buikema & Cairns, 1980).

The heading above has been taken directly from the title of a very significant contribution to this problem (Warren & Davis, 1971) which examines the whole rationale of artificial-stream techniques. One of the advantages of the laboratory-stream approach is that the reactions of individual species can be studied separately. Clearly, the more components of the fauna which can be examined in this way, the better the chance of interpreting crude field data and of beginning to analyse community reaction. In isolating various factors in order to study their effect on test organisms, it follows that in most cases the environment of the laboratory stream is a much simpler one than in nature. It follows logically that, from these simple (but essential) beginnings, the laboratory stream should progress gradually to more complex communities which are allowed to build up, along with the appropriate aquatic flora, in the artificial channel over a period, rather than be created overnight. It can now be appreciated that all of these ideas and conclusions have found expression somewhere in the development of techniques described earlier in this chapter. Unfortunately, the building up of such more complex stream communities requires additional resources which are not always available, and makes it progressively more difficult to arrange adequate replication or to carry out the large range of tests possible with simpler equipment. In order to make laboratory findings more readily translatable to field conditions and to reduce gradually the gap between the two phases of evaluation, those authors point out the need for laboratory investigation to be accompanied by experimental studies in natural field channels. There is no doubt that progress in the last 15 years since that critique was published has confirmed that many of these principles have been taken to heart. However, the gap between laboratory and field has not yet been satisfactorily bridged and we are not yet in a position to predict with any degree of certainty the likely outcome of stream contamination by a particular toxic chemical, and even less so when mixtures of toxicants are involved.

Many of these basic problems have been re-examined more recently in Symposia on Aquatic Toxicology and Hazard Assessment convened by the American Society for Testing and Materials (1982, 1983). There it was pointed out that among other factors likely to impair the value of laboratory tests is the quality of the dilution water (Chapman, 1983). Differences in the physicochemical characteristics of the laboratory dilution water and the body of water in question can be directly responsible for large differences in response of an organism to a toxic material, and to the physical and chemical form of the toxicant itself. Strict regard to this point has long been firmly

48

established in fish toxicology studies but has not yet taken its rightful place in stream invertebrate work. Possible objections on these lines can best be obviated by using water pumped straight from the natural stream to the laboratory channel. As has been described, this has been possible in some cases but not in others. Uniformity of life stage is also important for the achievement of consistent results. In addition, laboratory tests are usually carried out with test animals which have not been acclimatised to toxicant, and their response may well differ significantly from field populations which have been exposed to, and become acclimatised to, low sublethal concentrations over a long period.

The point is once more stressed that clear definition of the objectives of the laboratory tests, i.e. whether they are designed to establish tolerance levels of different species or whether they aim at predicting events in the field, can help to avoid misunderstanding and controversy. In conclusion, the author states that 'Much of the apparent inaccuracy of laboratory toxicity test data is attributable not to unnatural response of the laboratory test organism but to the extrapolation from a relatively narrow set of test conditions to a much broader range of environmental conditions' (Chapman, 1983).

In this connection it is important to remember that where there are apparent discrepancies between laboratory findings and field observations, the real possibility that the field data may be faulty or misleading due to sampling defects cannot be ruled out, and that both aspects need constant reappraisal.

Importance is now attached to the fact that while most laboratory tests involve exposure to a range of fixed concentrations of toxicant for a fixed period, this seldom occurs in nature where it is much more likely that exposure is cyclic or falls rapidly from a peak. The need to allow for this had already emerged from field studies on fenitrothion (Stanley & Trial, 1980), and the importance of extending laboratory tests to include a wider time/concentration profile has been amply confirmed in tests on *Simulium* larvae (Muirhead-Thomson, 1983).

There is now a great deal of information about the importance of chemical factors in determining how the action of a toxic chemical can alter according to varying composition in the natural habitat (Lee & Jones, 1983). Much of that information focusses on heavy metals and the responses of fish, but similar problems occur with many organic contaminants and have to be borne in mind in stream invertebrate investigations. Several degradation products of chlorinated hydrocarbon pesticides are more toxic to aquatic life than the parent compounds, such as dieldrin – a transform of aldrin – and heptachlor, more toxic than its parent heptachlor epoxide. It appears that one may have to consider whether the chemical used in the laboratory test – freshly diluted from a concentrate – is really identical to the complex of toxicant and degradation products created in the natural water body.

With regard to the controversial question of the relevance of laboratory data to field reality, this has been the subject of close scrutiny in the case of an aquatic invertebrate which normally inhabits still waters, and in which static laboratory tests have long been accepted usage for establishing acute and chronic toxicity data (Adams *et al.*, 1983). The test animal in question is the water flea, *Daphnia magna* which, because of its ease of handling and amenability to culture in the laboratory, has long been used (along with the fathead minnow. *Pimephales promelas*) to assess the safety of chemical effluents in aquatic environments and to develop water quality criteria, particularly in the US (see page 24). In this experiment, laboratory tests on exposure to toxic chemical, carried out by the conventional standard method for acute toxicity (24 and 48 h LC_{50}) and for chronic toxicity (21-day through-flow), were repeated exactly in outdoor simulated ponds 4000 l in capacity which were exposed prior to the test for a period of 6–8 weeks, sufficient to allow natural colonisation by pond flora and fauna. Despite the complexity of the simulated-pond environment, and the range of additional factors not present in the laboratory tests, a very close agreement was found between the two findings. The same concentration of chemical produced almost identical response in the laboratory and in the pilot-scale field test, and would appear to provide confirmation that in this instance laboratory tests do provide reasonable estimates of toxic effects on populations in natural aquatic ecosystems.

This type of crucial 'bridging test' should provide incentive to the carrying out of similar strict comparisons to be made between stream macro-invertebrate reaction in the laboratory simulated stream and in the much more exacting natural running water habitat. Until that time, while we would benefit from the guidance and example of the *Daphnia* experiment, there is no way we would be justified in concluding that the results with streams would necessarily show a similar degree of concordance.

The use of *Daphnia* has such a long and well-established history, and has been the subject of such intensive research from many directions, that its key role in establishing water-quality criteria and determining maximum allow-able toxicant concentration, has long been taken for granted. However, this established role has recently been reappraised (Buikema, Geiger & Lee, 1980) with the conclusion that the test methods are not as standardised or repro-ducible as we would like to believe. Many of the possible sources of error arise from the many variable factors involved in the laboratory culture of *Daphnia* with regard to nutrition, culture health and acclimatisation among others. When the time comes when the culture of stream invertebrates in the laboratory finally receives the serious attention which is long overdue, valuable lessons can be learned from these experiences with *Daphnia*.

As a replacement for *Daphnia* in running water studies, there is much to recommend in another crustacean, more typical of running water, and robust and amenable to long period of maintenance in the laboratory. This is the freshwater amphipod, *Gammarus*, in England the freshwater shrimp, mainly *Gammarus pulex*, and in the US mainly *G. pseudolimnaeus*, *G. fasciatus* and *G. lacustris*. In the 14th (1976) edition of the long established and authoritative US *Standard Methods* (American Public Health Association, 1976), tentative procedures for using amphipods (*Gammarus*) appeared for the first time. Since then, renewed interest in *Gammarus* has centred on its great potential for laboratory culture by standard methods and its great potential for bioassay tests (Arthur, 1980).

With regard to the European *Gammarus pulex*, a great deal of work has already been published by freshwater biologists about its physical and chemical requirements (Sutcliffe, 1971; Sutcliffe & Carrick 1973; Willoughby & Sutcliffe, 1976), and this species also played a key role in short-term toxicity studies, 1976–83, on stream macroinvertebrates in England (Muirhead-Thomson 1978 *a,b,c*, 1979). In pointing out the great possibilities of *Gammarus*, particular emphasis is placed on through-flow rather than static tests (Arthur, 1980). There is a whole new exciting field of investigation here for using *Gammarus* as the first test organisms to establish a closer link between laboratory data and field reality, and for using the ability of *Gammarus* to thrive in laboratory rapid through-flow aquaria, to establish a valid research programme on the cumulative and long-term effects on the organism of continued exposure to low concentrations of pesticide or allied toxic chemical.

FOUR

ARTIFICIAL COMMUNITY STREAMS
AND CHANNELS *IN SITU*

ARTIFICIAL STREAM COMMUNITIES

The laboratory streams described so far have been designed mainly for single species at a time, or for limited select groups of macroinvertebrates. The logical progress from this point is to design simulated streams in which impact of toxic chemical on a whole community can be studied under conditions akin to those in the natural habitat, but allowing certain factors to be controlled and studied separately in a way that is not feasible in the stream complex. Noteworthy developments along these lines have indeed been made, though not necessarily with the same objective. For example, in studies on the community effect of the lamprey larvicide TFM (see page 214) in Michigan, six fish hatchery channels 8 m long and 0.6 m wide were used, allowing three complete systems, each comprising one control and one adjacent experimental channel (Maki & Johnson, 1977). Each channel was divided into a 4 m upper pool section and a 4 m lower riffle section. The upper pool section was allowed to become colonised by introduction of organic matter and by drift of fauna in the gravity-fed water supply from an adjacent creek. The riffle section was also colonised from natural stream substrates, with associated fauna and flora introduced from a natural source. These communities were allowed to grow and become stabilised for a period of 2 months before experiments started, by which time a very good representation of stream organisms was established, including five species of stonefly (Plecoptera), three of mayflies (Ephemeroptera) and no fewer than nine species of caddis (Trichoptera), as well as the crustaceans *Gammarus* and *Asellus*.

The standing volume of a single stream, including pool and riffle sections, was approximately 550 l. The flow rate was 100 l/min per channel, the water being replaced completely every 5.5 min. The object of these experiments was to study the effect of the toxicant on the metabolism of benthic communities, changes in photosynthesis and respiration being measured by respiratory chambers installed in the substrate of the hatchery channels.

These experiments were able to demonstrate that respiratory measurements could provide a short-term assessment of the influence of the lampricide on a stream community. They also demonstrated the capacity of a stream ecosystem to adjust to the temporary effect of TFM; in the pool and riffle areas examined, the gross production, respiration and photosynthesis/respiration (P/R) ratios had returned to pre-treatment levels within 1–2 days following 1 complete day of exposure to TFM at the rate of 9.0 mg/l active TFM.

Even larger open-air artificial streams have been developed in Japanese studies on community effect of the organophosphorus compounds temephos (Abate) and chlorpyrifos (Yasuno, Sugaya & Iwakuma, 1985). The artificial model stream was 100 m long, 23 cm wide and a gradient of 2 %. The stream bed was covered uniformly with stones approximately 4 cm in diameter. Ground water was directed into the stream at the rate of 0.4 l/s producing a flow of approximately 12 cm/s. The flora and fauna were allowed to develop naturally from month to month until they reached a steady state suitable for experiments. At this stage, insecticide was applied to the head of the stream for a period of 30 min. Macrobenthos were sampled at several points in the channel, and drifting insects were trapped in a net at the channel outlet.

Unlike the Michigan channels described above, normal colonisation of these channels was slow. In one case, the fauna was dominantly chironomid, while in the other colonisation had proceeded a stage further to include species of caddis, hydropsychids. The effect of temephos was to destroy the chironomid population; this, in turn, was followed by increase in algal populations and the formation of an algal mat throughout the treated channel.

Introduction of chlorpyrifos produced a heavy drift of benthos on the part of both the chironomid *Thienemanniella* and the trichopteran hydropsychids, more than 100 000 of the former and 5700 of the latter being captured during the first 30 min.

The reactions of another chironomid, *Procladius* are of particular significance in the general context of unforeseen pitfalls in sampling and in interpretation of capture data. These larvae, which inhabited the middle section of the experimental stream where the water current was slower, were all killed by this insecticide, and a considerable number of dead individuals were found in this region. Only a few were captured in the drift net at the end of the stream, however, showing that in this case drift captures alone would have seriously underestimated both the lethal effect and the numbers affected.

The conditions in this very long artificial stream have clearly much in common with the large-scale stream established by the UK Freshwater Biological Association (see page 36), in both of which the early stages of colonisation were dominated by chironomid larvae (Pinder, 1985). The important difference remains however, that in the channel designed by the Japanese workers, water flowed straight through on a once-through basis, while the UK system is a very large-scale recirculating stream designed to study basic problems in community biology, and probably unsuited for experiments on impact of toxic chemicals.

An instructive example of the use of such outdoor community channels is provided by work on another organophosphorus insecticide, diazinon. Apart from the fact that with a high water solubility it was known to cause toxicant addition of 1–2% to freshwater ecosystems (Wauchope, 1978) very little critical work had been done on its impact on stream macroinvertebrates. Facilities at the ecological research centre involved, on the River Mississippi, included eight outdoor experimental channels, each 520 m long (Arthur *et al.*, 1983). The upper 122 m of three of these channels were used for evaluating experiments. The channels were supplied by a flow of Mississippi water at the rate of 0.8 m³/min. Each channel contained two riffles and one pool; a weir was used to raise the water level 15 cm at the downstream end of the first pool which acted as a sediment trap for incoming water.

The dominant macroinvertebrates colonising the channels were the amphipods *Crangonyx* and *Hyalella*, and the isopod *Asellus*, as well as smaller numbers of snails (*Physa*), flatworms (*Dugesia*), leeches and damselflies (*Enallagma* and *Ischnura*). Macroinvertebrate response to the toxicant was measured by community response, by drift changes and by insect emergence. In the dosing regime, two of the channels were dosed continuously for a period of 12 weeks, while the third was left as an untreated control. In regime (*a*), the two channels were treated at 0.3 and 3.0 µg/l (i.e. ppb) respectively. In regime (*b*), the two channels were treated at 6.0 and 12.0 µg/l of diazinon for 12 weeks, while in regime (*c*), the channel which had received the higher dosage in regime (*b*) was treated at 30 µg/l.

Benthic macroinvertebrates were sampled at the second riffle every 2 weeks, and a 30-h macroinvertebrate drift survey was conducted at the start of the diazinon dosing, with 30–48 h long surveys at the time of change from one dosing regime to another.

Sampling for water-column concentration profiles was conducted weekly in the pools at 2–5, 10–15 and 25–30 cms from the bottom and at the surface, and these records were able to provide a more accurate picture of the fate of diazinon in the channel. In regime (*a*), diazinon concentrations were maintained uniformly at 3.0 and 0.3 µg/l; but in regime (*b*), concentrations in the higher dosage channel – 12 µg/l – progressively decreased from the upper riffle through the pool to the lower riffle. A similar progressive reduction in measured diazinon concentration occurred in regime (*c*), with a reduction of about 25% from upper to lower riffles.

With regard to impact on the channel community, the diazinon treatment did not result in any consistent reduction in macroinvertebrate density, although increased drift – especially of amphipods and snails – occurred in treated channels 4 weeks after dosing began, and followed the increases in concentration regimes.

In evaluating the impact of a pesticide on aquatic invertebrates in static waters, there is no real barrier to designing tests which deal progressively with the intermediate phases between laboratory and field. This has been done on many occasions by the construction of small standard outdoor plots which provide a miniature ecosystem containing fauna and flora from a natural habitat. This method, which allows different pesticides, different formulations and different dosage rates to be compared under controlled laboratory-type conditions, has played an important part in translating laboratory findings to field realities in trials with mosquito larvicides, herbicides and others.

In the case of running water habitats, rivers in particular, a whole range of different factors provide apparently insuperable obstacles to coping effectively with this intermediate stage of evaluation. However, the stimulus of an urgent practical pest control problem provided the first real opportunity to bridge this gap, the urgent problem in this case being massive emergence of nuisance insects – mainly Trichoptera – from the St Lawrence River, Quebec, posing a potential threat to the success of the 1967 World Exhibition (Corbet, Schmid & Augustin, 1966). At that time, the most appropriate larvicides to deal with these aquatic pests were DDT and DDD in emulsifiable formulations (Fredeen, 1969).

Following the initial screening of larvicides in laboratory aquaria, it was decided to carry out multiple small-scale tests of selected larvicides in the St Lawrence itself, which at the time of these investigations had an average discharge of 6515580 l/s. This procedure not only allowed comparative tests to be carried out in near natural conditions, but also minimised the risk of polluting the river.

In a series of shallow rapids, where the water depth ranged from 1.1–1.22 m, larvicide-application points were sited side by side across the up-river edge of the general test area. Each single application point marked the apex of an individual test plot shaped like an elongated triangle. Within each test plot, two sampling stations for larvae were selected, one at 12.2 m and the other at 61 m downstream from the point of application. The pathway and width of spread of larvicide at each station was determined by using concentrated milk (from fat-free milk powder) as marker dye, and the width of the milky-coloured water was marked by small stone cairns or fluorescent orange plastic markers attached to rocks.

With measurements of depth, width and average water velocity – roughly in the 0.6–1.4 m/s range – controlled applications of larvicide, based on volume discharge, could be made with a high degree of accuracy. The controlled 30-min application was made from a 20 l polyethylene bottle suspended on a tripod a few feet above the river water; a constant discharge being achieved by means of a constant head.

The effect of treatment on aquatic organisms was judged by methods – drift nets and Surber samplers – which will be described more appropriately later. In assessing the value of such miniature tests, it was pointed out that there were certain discrepancies and errors due to the fact that the margins of the treated area were based on visual judgement of the marker-dye pattern, and that in future trials of this type the accuracy could be improved by chemical analysis to determine exact larvicide concentrations within the test plot.

Those trials in the St Lawrence River, prompted by an unusual pest control system, have not been followed up to any significant extent in Canada or the US, and remain a relatively isolated example of this trial *in situ* in that region. However, they provide the forerunner for what is now regarded as one of the key evaluation techniques in another much vaster and longer lasting control project directed against undesirable aquatic fauna, namely, the Onchocerciasis Control Programme (OCP) in West Africa for the control of *Simulium damnosum* populations in the Volta River Basin and associated river systems (see Chapter 5).

The French school of freshwater biologists in ORSTOM (*Office de la Recherche Scientifique et Technique Outre-Mer*) were mainly responsible for aquatic research leading up to the official beginning of the programme in 1974, and have continued to play a major role ever since. They were very conscious of the fact that at that time knowledge about the collection of live field material, and the handling and maintenance of sensitive stream organisms – including *Simulium* larvae – was still in an early exploratory stage, and that these organisms were liable to show a degree of trauma or mortality incompatible with valid laboratory testing. In view of this they considered that the laboratory tests had doubtful relevance to natural habitats, and they therefore concentrated their attention on designing tests *in situ* which would represent a reduced model of a river bed, but provide the same accuracy as in the laboratory (Dejoux, 1975).

The basic element of the equipment is a robust and portable galvanised iron experimental channel (*une gouttière*). The apparatus is set up in a shallow part of the river in line with the current, kept in position with iron supports which also regulate the height above the natural stream bed, and furnished with substrates from the river bed populated with natural invertebrate fauna. The

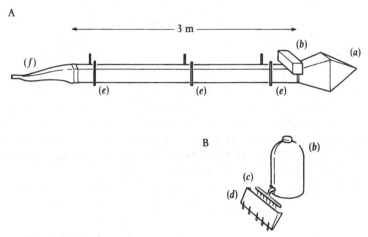

Fig. 4.1. A, channel *in situ* (*gouttière*) used in experimental studies on impact of *Simulium* larvicides in the OCP, West Africa (after Dejoux, 1978). B, Alternative arrangement of larvicide reservoir and dispersion component (after Dejoux, 1982). (*a*) Net to prevent ingress of natural drift, (*b*) insecticide reservoir, (*c*) distribution ramp for larvicide, (*d*) chemical dilution mixer, (*e*) stabilising and height-regulating struts, (*f*) drift net.

lower portion is fitted with a removable drift net, and over the upper part of the channel is a storage tank for the pesticide formulation which can flow regularly into the channel.

The design of experimental channels now used *in situ* differs slightly from the prototype and is shown in Fig. 4.1 (Dejoux 1978). It is 3 m long and from 15 to 20 cm wide. Each test is carried out as follows. Substrates with stream fauna are installed in the channel and allowed to settle and redistribute over a period of at least 6 h, during which time the channel is open at both ends with the stream flowing through. The drift net is then installed at the lower end, and the insecticide formulation allowed to flow into the upper end at a discharge calculated from the current velocity, and according to the desired concentration and time of exposure.

The larvicide flowing into the channel is distributed through several outlets extending across the width of the channel; this, and an array of baffles in the upper part of the channel ensures that the fauna in the channel are exposed to a uniform flow of chemical. Drift net collections are removed 30 min after application time; nets are quickly replaced, and further collections recorded at 1, 2, 4, 6, 8 and 24 h after initial treatment. At the end of that period, the two ends of the apparatus are closed and the entire assortment of substrates is removed from the channel for sorting and examination of resident fauna.

The information obtained from each test is: number and drift rate of different organisms during the 24 h after treatment, and the speed of response of different species to the passage of the larvicide. This enables an overall drift pattern (*cinétique du déchrochement*) to be determined for each species, as well as the relative tolerance of different species to the chemical. By using two or three channels simultaneously – one of them being a control – it has been possible to make accurate comparisons between different formulations of Abate (temephos), the main larvicide in the OCP, and to compare the effect of Abate with that of other larvicides under consideration (Dejoux & Guillet, 1980), particularly chlorphoxim (Troubat, 1981; Dejoux *et al.*, 1982) and permethrin (OCP, 1985 *a,b,c*).

These channels can also be used to monitor the effects of practical control measures carried out by aerial spraying on particular rivers or sections of rivers by installing them 24 h before the spray schedule, and providing them with the requisite populations of aquatic invertebrates. This technique has played a key role in the evaluation of *Simulium* control in the OCP and in a comparison of the environmental impact on non-targets of different formulations and candidate chemicals. The results obtained have to be assessed in the context of the whole programme and in conjunction with information obtained by other more conventional techniques also used.

The experimental channel *in situ* has provided a valuable insight into the effects of Abate and other *Simulium* larvicides on the drift patterns of both target and non-target invertebrates. However, the French scientists in its development and practical use have been very conscious that there is some information about drift organisms which the apparatus in its conventional form could not provide. This concerns the question of whether or not a significant proportion of drift organisms can eventually reattach to a substrate and survive. As the net at the downstream end of the 3 m channel intercepts the drift, it is impossible to say what would have happened if the drift had been allowed to continue uninterrupted, as it would be in a normal treated river.

In order to throw light on this question, a modified design of channel using the same basic principle was devised (Dejoux, 1982). Two of the conventional 30 m long zinc channels were attached together by means of a mobile plastic sleeve so that they could form a single 60 m long channel. A few days before the experimental treatment was to be carried out, the channels were installed partly submerged above the river bed. The upstream channel was furnished with natural substrates and introduced fauna as previously. The downstream channel was left untouched. On the day of the experiment the lower channel was swept clean of any debris or fauna which might have collected, and this channel was then closed at both ends. Various types of natural substrate from the stream were then installed in the lower channel after they had first been completely cleared of all attached fauna. The substrates were arranged in

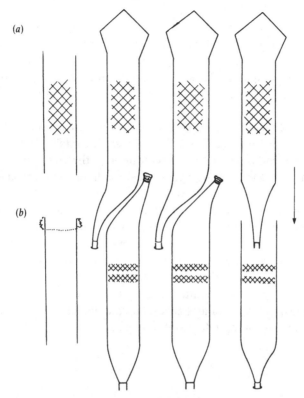

Fig. 4.2. Arrangement of experimental channels *in situ* in studies on re-attachment of drift organisms in insecticide-free zone of channel. (*a*) Upper channels, (*b*) lower channels (after Dejoux, 1982).

such a way as to create three zones in the channel; at its upper end there was a slow-flowing stretch of about 1 m; this was followed by a stony zone of about 50 cm in which the current flow could range from 20 up to 60 or 80 cm/s according to the exact site. This riffle or 'rapids' zone was followed by a 1.5 m stretch of slow-flowing water at the downstream end of the channel. The object of all this was to provide a natural range of substrates and flow conditions to meet the requirements of different organisms.

The actual procedure of testing is illustrated in Fig. 4.2 (see also Dejoux, 1982). Before introducing Abate into the upper channel in the manner previously described at the rate of 1.0 ppm for 10 min, the lower channel is disconnected and both ends closed with netting. In the first 30 min after application, organisms drifting in the upper channel are collected in the net at the lower end of that channel, collections being made every 10 min. It is in this first half hour that drift mortality is at its highest, and includes a high proportion of early instars of sensitive invertebrate groups. At the end of that

30 min period, the drift net at the lower end of the upstream channel is removed, and the two channels are connected by the plastic sleeve to produce a single long channel. From then onwards drift organisms from the treated channel can be carried into the lower uncontaminated channel without any obstacles or interruption. Those organisms which do not re-attach, but continue to drift, are collected in nets at the lower end of the downstream channel.

For the next 2 h collections from the drift net are made every 10 min, and thereafter at longer intervals. The drift organisms are retained for observation *in situ* in plastic containers provided with netting, through which there is a flow of clean river water. Their physiological condition, is recorded from time to time up to 24 h and, at the end of that time, the substrates in the lower channel are examined and a record made of all attached organisms.

This equipment provided information about the drift patterns of different invertebrate groups or species in the course of the 24 h following the 10-min injection of larvicide. It also provided a record of the mortality of organisms in the different drift fractions, and finally it shows that components of the total drift are capable of re-attaching and surviving in the 24 h after treatment. Further details of actual results obtained in the OCP studies will be described in the appropriate context (page 201).

FIVE

INVERTEBRATE REACTION TO PESTICIDES

UNDER LABORATORY AND EXPERIMENTAL

CONDITIONS

TOLERANCE LEVELS UNDER STANDARD CONDITIONS

REACTIONS IN STATIC TESTS

With the widespread use of DDT and allied chlorinated hydrocarbon insecticides for pest control in the 1950s and 1960s it became increasingly clear that stream invertebrates were highly vulnerable to these chemicals. This destructive effect was particularly evident when stream and river habitats were unavoidably contaminated by repeated aerial applications of insecticides against forest pests such as the American spruce budworm. Experiences with mass destruction of river fauna in one of these campaigns in the Yellowstone National Park in the US, stimulated the need for more precise information about the susceptibility of stream invertebrates to the different insecticides then in use. This resulted in perhaps the first serious development of a laboratory evaluation programme for stream invertebrates in general as distinct from such target fauna as *Simulium* larvae (Gaufin, Jensen & Nelson, 1961; Jensen & Gaufin, 1964, 1966; Gaufin *et al.*, 1965). That and the other contemporary work has already been reviewed in depth (Muirhead-Thomson, 1971), but there are still aspects of those studies which are of particular significance in the light of developments in the 20–25 years since that period. First of all, two aquatic invertebrates which played a prominent part in those laboratory tests, namely the stonefly (Plecoptera) nymphs of *Pteronarcys californica* and *Acroneuria pacifica*, were typical of clean unpolluted running water and were also important food organisms of trout. Second, both species were robust creatures, easy to obtain and to maintain in a healthy condition in the laboratory. In this respect they compared favourably with some other fauna typical of the same habitat, e.g. the caddis *Hydropsyche*, which was much more difficult to maintain alive and thus less well suited for critical tests. Third, although the first series of laboratory tests were carried out in

Table 5.1. *96-h LC$_{50}$ values for two species of stonefly naiads arranged in decreasing order of toxicity (ppm)*

	Acroneuria pacifica		Pteronarcys california
Endrin	0.00039	Endrin	0.0024
Parathion	0.00280	Guthion	0.0220
Malathion	0.00700	Parathion	0.0320
Guthion	0.00850	Dieldrin	0.0390
Dieldrin	0.02400	Malathion	0.0500
Aldrin	0.14300	Aldrin	0.1800
DDT	0.32000	DDT	1.8000

After Jensen & Gaufin, 1964.

static water, at standard 24-h exposure periods, the authors were early to recognise the limitations of static tests for organisms which live in flowing water, and to modify these test vessels to permit a continuous through-flow of water.

That work also established two guiding principles, fully endorsed by subsequent work over the years, namely that the tolerance levels of a particular species may differ widely according to the nature of the pesticide. In their series they found that the chlorinated hydrocarbon insecticide endrin emerged as the most toxic to both species, with DDT the least toxic, while several organophosphorus compounds occupied an intermediate position (Table 5.1). The second point was that closely related genera sharing the same habitat may show consistent differences in reaction to one and the same chemical.

There is now a great deal of information both from the laboratory and from field experimental plots about the tolerance levels of different aquatic invertebrates to a wide range of pesticides in general and insecticides in particular (NTIS, 1972). In the context of this book, there would be little real advantage in cataloguing this mass of information from many different sources; the assessment of all this data can only be made with full regard to the test conditions which vary so much from project to project and which in so many cases are based on static tests.

REACTIONS UNDER FLOW-THROUGH TEST CONDITIONS

With increasing recognition of the need for laboratory tests on stream invertebrates to be based on systems which employ water flow or rapid water change, the main backdrop to this particular review will be the continuous laboratory studies on stream invertebrates carried out from 1973 to 1983, first on stream fauna in southern Africa (Zimbabwe) and latterly in the UK (Muirhead-Thomson, 1973–83). One technique which was used continuously throughout that period was the rapid through-flow chamber, with standard exposures of 1 h to various pesticide chemicals, followed by 24 h in a flow of clean water (see page 33). For further information on drift reactions or other behavioural effect, shorter exposures of 15 and 30 min were used in a miniature simulated laboratory stream (page 41). With such reasonably standardised techniques suitable for a wide range of stream invertebrates, both tropical and temperate, it has been possible to build up comparative and consistent data about tolerance levels of select stream invertebrates to the range of insecticides which have played a major role in the last 15 years.

The first of this continuous series was carried out on invertebrates from small streams in Zimbabwe. Apart from *Simulium*, baetids and hydropsychids, there was a paucity of stream forms which rather restricted the scope of the project. This was partly offset by the abundance of 'torrential dragonflies' – one agrionid *Pseudagrion kersteni*, and three libellulids, *Crocothemis divisa, C. sanguinolenta* and *Zygonyx torrida*, enabling that group to be laboratory-tested for the first time. It is those tests with dragonfly naiads which produced the most striking results in that all four species showed a very high degree of tolerance to DDT, and – to a slightly lesser degree – to another chlorinated hydrocarbon insecticide Thiodan or endosulphan. No mortality resulted from 1 h exposures to concentrations of DDT up to 2 ppm, and 80–88% survived the same length of exposure to very high concentrations of 10 ppm and 20 ppm respectively. This high tolerance contrasted strikingly with the high sensitivity of *Simulium* and *Baetis* under identical test conditions, both of these organisms showing 95% mortality after a 1-h exposure to 0.1 ppm DDT (Muirhead-Thomson, 1973).

In the case of the dragonfly test species, the high number surviving after the 24-h holding period in clean water were maintained for a further holding period when the completeness of their survival was confirmed when they eventually produced adults. By increasing the length of exposure period to the two highest concentrations it was shown that about 50% of the dragonfly naiads still survived a 4-h exposure to 20 ppm, and 16% survived a 24-h exposure to 10 ppm. Again, in both cases, survivors could be maintained in the laboratory long enough for adults to emerge.

In contrast to their high tolerance to DDT, all four species of dragonfly naiad were highly susceptible to the organophosphorus compounds fenthion (Lebaycid, Baytex) and parathion (Folidol), showing the same degree of sensitivity as *Simulium* and *Baetis*, with exposures of 1-h to concentrations of 0.05 ppm upwards producing nearly 100% mortality.

In the series of trials with European fauna, the same procedure was followed as before, with a 1-h exposure in rapid through-flow, followed by a 24-hour holding period in a continuous flow of clean water. In accordance with the differences between the running water habitats and faunal requirements in the two regions, the European tests were carried out at laboratory temperatures of 17.5 °C $\pm$ 1.0, while those in southern Africa had been done at 21–22°C.

In the first of the European test series a comparison was made between temephos (Abate) – the main larvicide used in the West African *Simulium* control programme (OCP, see page 189) – and another organophosphorus insecticide chlorpyrifos methyl, at that time already under consideration as an alternative or successor to Abate (Muirhead-Thomson, 1978). The range of macroinvertebrates tested was as follows.

Crustacea	Amphipoda	*Gammarus pulex*
Insecta	Ephemeroptera	*Baetis rhodani*
		Ephemera danica
	Trichoptera	*Brachycentrus subnubilis*
		Hydropsyche pellucidula
		Rhyacophila dorsalis
	Odonata	*Agrion splendens*
	Diptera	*Simulium ornatum*
		S. equuinum
		S. erythrocephalum
	Rhagioniidae	*Atherix* sp

The results are summarised in Table 5.2 and show the relative susceptibility of the six indicator stream invertebrates which formed the bulk of the test material. The results show that the mayfly nymphs of *Baetis* are the most highly susceptible organisms to both chemicals. This sensitivity had previously been recorded to a range of organochlorine and organophosphorus compounds in the African tests (Muirhead-Thomson, 1973) and in preliminary tests with European species (Muirhead-Thomson, 1971), but this was the first series extensive enough to determine the lower threshold concentrations and to show that it was even more susceptible to both chemicals than the target organisms, *Simulium* larvae. The figures indicate that *Baetis* is about 10 times more sensitive than *Simulium* to chlorpyrifos, and about 100 times more sensitive in the case of temephos.

The high sensitivity of *Baetis* to these compounds is by no means typical of Ephemeroptera as a whole. A short series of tests with *Ephemera danica*, *Ecdyonurus* sp. and *Caenis* sp. all indicated much higher tolerance levels to these chemicals.

Perhaps the most unexpected result was the wide difference in the reactions of the amphipod *Gammarus pulex* to the two chemicals, the unusually high tolerance to temephos (Abate) contrasting with its very high sensitivity – only exceeded by that of *Baetis* – to chlorpyrifos. As 100% of the

Table 5.2. *Relative susceptibility, in increasing order, of stream macroinvertebrates to temephos and chlorpyrifos as judged by estimates of the minimum lethal concentration (i.e. the lowest concentration to produce 90–5% mortality 24 h after a 1-h exposure) at 17.5 °C*

Temephos (ppm)		Chlorpyrifos (ppm)	
Baetis	0.001–0.002	*Baetis*	0.01–0.02
Agrion	0.04–0.05	*Gammarus*	0.05–0.1
Brachycentrus	0.1–0.2	*Simulium*	0.05–0.01
Simulium	0.2–0.5	*Agrion*	0.2
Hydropsyche	0.5–1.0	*Brachycentrus*	0.2–0.5
Gammarus	≫ 1.0	*Hydropsyche*	> 0.5

After Muirhead-Thomson, 1978.

Gammarus consistently survived the highest concentration of temephos used, 1 ppm, there was no opportunity for defining the minimum lethal concentration, but it must have been of the order of many thousand times greater than that of the most susceptible organism, *Baetis*.

This sharp difference in reaction to these two organophosphorus insecticides on the part of amphipods related to *Gammarus* also emerged clearly from studies carried out in southern California where these chemicals were being tested to control chironomid larvae in residential recreational lakes (Ali & Mulla, 1977). The amphipod in question, *Hyalella azteca* or 'side-swimmer' was quite unaffected by treatment of those habitats at the rate of 0.17–0.28 kg AI/ha. In similar rates of treatment of these chironomid habitats with chlorpyrifos which produced concentrations of 0.0074 ppm, *Hyalella* was severely affected, and completely eliminated for 3 weeks following treatment.

In an earlier report on the high tolerance of *Hyalella azteca* to temephos, an LC_{50} value in the laboratory as high as 2.5 ppm had been reported (Van Windeguth & Patterson, 1966)

Observations on microcrustacea have also shown wide differences in impact between these two chemicals, at least on the fauna of static habitats. Abate treatment, for example, was found to reduce or eliminate *Daphnia* spp (Cladocera) but had no effect on the copepods, *Diaptomus* and *Cyprinotus*. Chlorpyrifos had no adverse effect on the copepods *Diaptomus* and *Cyclops*, but nearly eliminated *Cyprinotus* and also markedly reduced *Daphnia* (Ali & Mulla, 1977).

Concurrent with the studies on the European fauna using 1 h exposures in rapid through-flow test vessels, additional tests were being carried out in

65

the miniature simulated stream (page 41) in order to determine, (a) the effect of different exposure periods, and (b) drift patterns. These will be discussed later.

In a further series of 1 h exposure tests under rapid through-flow conditions, attention was given to another organophosphorus insecticide, chlorphoxim, and to two synthetic pyrethroids, permethrin and decamethrin, all of which were under consideration as possible replacements for Abate (temephos) in the OCP *Simulium* control programme in West Africa (Muirhead-Thomson, 1981 a,b,)

Of all these compounds, decamethrin proved to be by far the most lethal to all macroinvertebrates tested. This was not unexpected in view of earlier reports on its very high toxicity to mosquito larvae (Mulla, Navvab-Gojrati & Darwazeh, 1978; Mulla, Darwazeh & Dhillon, 1980).

On the basis of LC_{95} figures, which correspond closely to the minimum lethal concentration, decamethrin emerged as about 20 times more toxic than Abate to *Simulium* larvae, and about 4 times more toxic than chlorphoxim. All of the non-target invertebrates were affected by roughly the same extent, the most susceptible organism being once more the nymphs of *Baetis rhodani*. After a 1-h exposure followed by a 24-h holding period, the minimum lethal concentration was 0.05 ppb (μg/l), and mortalities over 50% were consistently recorded at the lowest concentrations of decamethrin tested, viz, 0.01 and 0.005 ppb.

Only one organism tested, the mayfly *Ephemerella ignita*, appeared to be less susceptible to decamethrin, over 60% surviving a concentration – 0.5 ppb – which produced mortalities at or near 100% in all the other test organisms.

The tests with chlorphoxim also produced results of considerable interest in that it was the first insecticide tested in this series to which *Simulium* larvae were significantly more susceptible than any of six non-target species. This is best illustrated when the results of all tests in the 0.002–0.005 ppm (i.e. 2–5 ppb) experiments are combined and listed in order of decreasing susceptibility (Table 5.3).

With regard to the unusually high toxicity of the synthetic pyrethroid decamethrin experienced in that test series, it would be useful to consider other workers' experiences with these toxic chemicals. The high toxicity of decamethrin to mosquito larvae in static water habitats was established in mosquito control investigations in the US (Mulla *et al.* 1978, 1980). The results of laboratory tests carried out under static conditions and based on 24-h exposures are not directly comparable to the 1-h exposures used in the rapid through-flow of the European series, but they do indicate susceptibility levels of the same low order, ranging from LC_{90} figures of 1 ppb for the least susceptible species to 0.04–0.08 ppb for the most susceptible.

A considerable amount of work has been carried out recently on another synthetic pyrethroid, cypermethrin, the active ingredient of many insecticides

Table 5.3. *Summary of mortalities produced by chlorphoxim (24 hours after a 1 h exposure to concentrations of 0.002–0.005 ppm) in 7 stream macroinvertebrates, in decreasing order of susceptibility.*

	Number of tests	Total tested	(%) Mortality
Simulium	15	4609	93
Baetis	18	1042	70
Gammarus	9	393	46
Rhyacophila	11	116	25
Ephemerella	2	48	25
Hydropsyche	14	127	12
Agrion	7	26	0

After Muirhead-Thomson 1981*a, b*.

widely used in crop protection (Stephenson, 1982). Among the many test animals used in 24-h static toxicity tests were two typical stream organisms, *Gammarus pulex* and the ephemeropteran *Cloeon dipterum*, whose LC_{50} values were 0.1 and 0.6 μg/l (ppb) respectively. The high sensitivity of *Gammarus* was used to develop a bioassay technique, the word 'bioassay' being used here in its strict sense as the use of living organisms to detect the presence, or concentration, of a toxic substance. In the technique developed in studies on contamination of freshwater ditches resulting from aerial application of cypermethrin (Shires & Bennett, 1985) water samples from these ditches were routinely taken to the laboratory and divided among three dishes, to each of which 10 *Gammarus* were added; mortalities were noted 24 h later. These tests showed that no significant mortality occurred beyond the first 2 days after spraying, and it was concluded that such a bioassay, based on sensitive indicator species, was perhaps a better indication of the presence of biologically significant cypermethrin residues than are chemical analyses carried out in waters containing such low concentrations.

TIME/CONCENTRATION IMPACT

The use of a standardised laboratory test for establishing susceptibility levels for stream invertebrates under flowing water conditions, as distinct from static, is an essential first step in the overall evaluation of pesticide impact on stream fauna. Provided control mortalities are maintained at an acceptable low level during exposure and recovery period, and the results are consistent from test to test, such a test series can provide the first indication of major differences in the response of different organisms to the same chemical, or the response of the same organisms to different chemicals.

Laboratory and experimental methods in evaluation

However, this is only a first step; the exposure period selected for the series just described, 1-h, is an arbitrary and convenient one as also is the 24-h recovery period – which may be inadequate for some slow-acting toxicants (Lacey & Mulla, 1977 *a,b*). As stream organisms under pesticide impact in their natural habitat are liable to be exposed, not only to a wide range of concentrations but also for varying periods of time, the relevance of such standardised laboratory findings to actual field conditions is therefore limited. Nor can one extrapolate from the 1-h experimental data on the assumption that – even within a restricted range – concentration × time equals a constant. This question of time/concentration impact has long been one of concern, for example, in the practical field of evaluating larvicides for *Simulium* control (Jamnback & Means, 1966; Jamnback, 1969) but the issues have been rather clouded by the fact that in early trough experiments the detachment of *Simulium* larvae when exposed to chemical was automatically taken as a measure of mortality.

As mentioned in the previous section, the long series of laboratory tests based on a 1-h exposure in a rapid through-flow test vessel was accompanied by a series using the miniature simulated stream; the advantage of the latter being that shorter exposures of 30, 15 or 5 min could be performed much more precisely than in the through-flow technique. And of course it had the added advantage of enabling drift patterns to be determined. Comparison of data from the two different techniques, with particular reference to *Simulium* larvae, disclosed an apparent anomaly in that the minimum lethal concentration for Abate obtained in the 1-h flow-through test was very much lower than would have been expected from extrapolation of data provided by 15-min exposures in the simulated stream. As this apparent discrepancy might have been due to some undetermined difference in the organism's reactions to the two different techniques, it was clear that any valid test on the impact of different time/concentration regimens must be based on an absolute uniformity of test procedure. In addition, it was an essential requirement for such a strict comparison that control mortalities must be maintained as close to zero as possible. As the miniature simulated stream was best suited to fulfil these requirements for the particular test organisms in question (*Simulium* larvae), it was the technique of choice for comparing the effect of three different exposure periods, 15 min, 1 h and 2 h (Muirhead-Thomson, 1983).

Simulium larval collections brought in from the field were distributed among four identical miniature simulated streams. In two of these larvae were exposed for the longer period of 1 h in one series and 2 h in another series, while the larvae in the third channel were exposed for 15 min. The fourth channel was the untreated control. In each of these weekly experiments, time of exposure x concentration (Abate) was the same, 0.1 ppm for 15 min being matched by 0.025 ppm for 1 h, and so on. In all of these tests

68

68

Table 5.4. *Summarised results of 22 experiments on the effect of different time/concentrations of Abate (Procida 200 EC) on late instar larvae of Simulium ornatum and S. equinum as judged by percentage mortality after 24 h in clean water. Exposures in laboratory miniature simulated stream at 17 ± 1 °C. (Overall composition of samples, S. ornatum 58%, S. equinum 42%.[a])*

	Periods of exposure		
	15 min	1 h	2 h
Concentration of Abate	0.4 ppm	0.1 ppm	0.05 ppm
Total larvae	487	1527	239
Number of tests	3	4	2
Mortality (%)	30	93	99
Concentration	0.2 ppm	0.05 ppm	0.025 ppm
Total larvae	717	1094	278
Number of tests	2	3	2
Mortality (%)	1.4	31	100
Concentration	0.1 ppm	0.025 ppm	0.0125 ppm
Total larvae	988	1143	535
Number of tests	2	2	2
Mortality (%)	6.6	34	94

[a] Controls: 9 experiments, 4575 larvae, 0.36% mortality (range 0–1.4%). (After Muirhead-Thomson, 1983.)

freshly made-up dilutions of the Abate E.C. formulation were run through the experimental channels at the rate of 80 l/h.

The results obtained are sufficiently noteworthy to be recorded fully (Table 5.4) and they show that over a wide range of Abate concentrations the longer exposures of 1 h and 2 h produced a much higher mortality than the short 15-min exposure to a correspondingly higher concentration. This is particularly marked in the 0.1 ppm for 15-min series which resulted in a mortality of 6.6% as compared to mortalities of 34% and 94% produced by the same absolute quantity of Abate applied for 1 h (0.025 ppm) and 2 h (0.0125 ppm) respectively.

Exposures to Abate concentrations of 0.1 ppm for 15 min are of particular significance in that this is a dosage rate very close to that widely used in practical *Simulium* control in streams and rivers viz., 0.1 ppm for 10 min. As the wave of larvicide becomes attenuated in its passage downstream from the application point, stream organisms become exposed to the chemical wave for progressively longer periods, but at correspondingly decreasing concentrations. The implication from these preliminary laboratory trials is that the lethal and controlling effect on *Simulium* larvae may not be due so much to the practical field application rate (0.1 and 0.05 ppm for 10 min) but to the increasingly longer exposure to decreasing concentrations as the larvicide moves downstream.

Unfortunately, no other 'non-target' fauna have been investigated in this series as thoroughly as *Simulium* with regard to time/concentration impact, but there is some preliminary evidence that some of them may react in the same manner. For example in tests carried out on *Gammarus pulex* at 1-h exposures to chlorphoxim – tests not sufficiently complete to establish LC_{90} levels – a concentration of 0.005 ppm produced 86% mortality. The corresponding time/concentration for a 15-min exposure, if one assumed that time × concentration = a constant, would be 0.02 ppm; but in fact a concentration of 0.05–0.06 ppm was required to produce the same mortality (Muirhead-Thomson, 1981 *b*). A similar relationship emerged from parallel tests on reactions of *Baetis* to chlorphoxim, where a 1-h exposure to concentrations of 0.005 ppm upwards consistently produced 100% mortality. The time/concentration equivalent for a 15 min exposure would be 0.02 ppm, but the real experiment produced a mortality of only 24%.

The trend of these figures does lend support to the idea that other stream invertebrates besides *Simulium* may also be more susceptible to longer exposures to low concentrations than to proportionately short exposure to the higher concentration, at least in the case of the two organophosphorus insecticides mentioned. If so, it adds an important new dimension to the evaluation of pesticide impact in running waters. The implications will be further discussed later in the light of the increasing amount of information now available from physicochemical assay about pesticide concentrations recorded in running waters exposed to contamination. Like so many evaluation techniques specially designed for running water macroinvertebrates, the results of this series of time/concentration tests can only be regarded as preliminary or exploratory, but they point clearly to an obvious line for further research, under similar laboratory controlled conditions, namely the reactions of organisms at much longer exposures to reputedly 'sublethal' concentrations. Progressively increased exposures of 24, 48 and 72 h have long been practised in static water tests, but – until very recently – there has been very little information about the effects of such exposures under conditions of *rapid* through-flow or of simulated stream. This may possibly be because of the much greater time and labour involved in carrying out an acceptable number of tests and replicates, many at such low part per billion concentrations that test solutions must be prepared manually and not left to automatic serial-dilution equipment.

An earlier attempt was made to provide much needed information on this aspect on invertebrate reaction using 'torrential dragonfly' naiad fauna of an African stream (Muirhead-Thomson, 1973). One of the distinctive reactions of all four species involved was their consistently high tolerance to DDT – see page 63 – in 1-h through-flow tests supported by further experiments at 4-h and 24-h exposures. In contrast all dragonfly species were highly susceptible to the organophosphorus insecticide fenthion (Baytex). Their robust naiads were well suited to longer-term experiments as the controls consistently registered zero mortality over several days. With regard to DDT, the 'sublethal' concentration selected – at which 100% of the organisms had survived the 1-h exposure – was 1 ppm. When exposures were extended to 48 h and 72 h under identical rapid through-flow conditions, viz., 45 l/h, the results obtained were as follows:

Mortalities after different exposures (%)

	1 h	48 h	72 h
Agrionid dragonflies	0	68	68
Libellulid dragonflies	0	90	87

These results indicate two aspects of prolonged exposure, firstly that, with an organism with a high tolerance level to DDT, increasing exposure does reveal a point at which a lethal effect appears; and secondly that, beyond a certain point, further increases in duration of exposure may not necessarily be followed by corresponding increases in mortality. Another interesting trend was that the prolonged exposure revealed differences in susceptibility of the two groups of dragonfly nymphs, which were not evident at the 1 h exposure level. These responses may be peculiar to these particular test organisms, or to the chemical action of DDT itself, or perhaps to both causes, but these observations do introduce one more facet into the complexities of evaluation.

Experiences from many different sources on the laboratory testing of stream invertebrates has shown that some organisms are unusually robust and can be maintained in the laboratory for long periods in a healthy condition, while others are less adaptable and begin to show increasing mortality after a few days, even in simulated-stream conditions.

Several of these robust indicator species have been the subject of one of the few laboratory investigations involving continuous exposure to pesticide in flowing water conditions for periods up to 28 days (Anderson & DeFoe, 1980). The test equipment used (DeFoe, 1975) allows a through-flow in each exposure chamber at the rate of 22 l/h. The two pesticides selected were endrin and methoxychlor; the test animals used for the former being nymphs of the caddis fly *Brachycentrus americanus* and the stonefly *Pteronarcys dorsata*, and for the latter the caddis fly *Hydropsyche*, the stonefly *Pteronarcys*, the isopod *Asellus communis* and the snail *Physa integra*. Test animals were exposed continuously to 5 different concentrations for 28 days, during that time all test animals being observed for toxic or behavioural effects.

In the case of methoxychlor the main interest is in the reactions of *Asellus* and *Hydropsyche* – stoneflies and snails being unaffected by the 28-day exposure. The reactions of *Asellus*, the most sensitive species tested, were

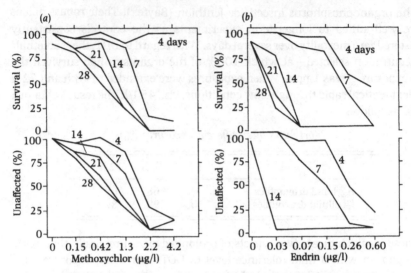

Fig. 5.1. (*a*) Effects of methoxychlor on survival (top) and normal behaviour (bottom) of *Asellus communis* after exposure for 4–28 days. (*b*) Effects of endrin on survival (top) and normal behaviour (bottom) of *Brachycentrus americanus* after exposure for 4–28 days (after Anderson & DeFoe, 1980).

recorded in terms of behaviour, EC_{50}, and survival, LC_{50} (Fig. 5.1*a*). After the first 4 days the number of deaths increased with increased exposure time, the LC_{50} falling from 1.75 ppb at 7 days to 0.42 at the end of 28-day observation period. In the case of *Hydropsyche* there was no sign of toxic action to the range of concentrations during the first week, but from the second to the fourth week a change in toxicity was shown by the lowering of the LC_{50} values from 2.9 ppb to 1.03 ppb. With endrin, the caddis fly *Brachycentrus* was the most sensitive animal tested, toxic effects being noted after the second day at the highest concentration (0.6 ppb). The results (Fig. 5.1*b*) show that after 14 days the only survivors were at the lowest concentration of 0.03 µg/l (ppb), so that an LC_{50} could not be calculated. Behavioural changes which occurred within the first 4 days took the form of case-leaving by the larvae.

This extended flowing-water test is a clear demonstration of the fact that as the time of exposure increases, the toxic effects are seen at lower concentrations. They also reveal differences in sensitivity – at least to methoxychlor – on the part of two closely related caddis larvae, viz., *Hydropsyche* and *Cheumatopsyche*. In the former the LC_{50} for a 4-weeks exposure was 1.30 µg/l (Anderson & DeFoe, 1980), while in the more sensitive *Cheumatopsyche* the figure was 0.42 µg/l (Merna & Eisele, 1973). In this series of tests, investigations were also carried out on the bioaccumulation of these two chemicals; this will be discussed later.

Further information about the comparative effects of short-term and

72

long-term exposures has been provided by a study on the reactions of three non-target macroinvertebrates to Flucythrinate, a synthetic pyrethroid, namely *Gammarus lacustris, Pteronarcys dorsata* and *Brachycentrus americanus* (Anderson & Shubat, 1984). The apparatus used consisted of a serial dilution system supplying 12 flow-through test chambers at the rate of 28 ml/min, 5 concentrations being tested at a time. Two types of experiment were carried out. The first of these, involving continuous exposures, took two forms: (*a*) observations on the effects of 5 concentrations were made at 1, 3, 6, 12, 24, 36, 48, 72 and 96 h and (*b*) observations on continuous exposure for 28 days. In the second type, short exposures were carried out for specific time periods from 0.5–96 h, after which the test animals were transferred to clean water.

In the first test, LC_{50} calculations for *Gammarus* were made for 48 h, 72 h and 96 h exposures, and these showed that with continuous exposure the LC_{50} values decreased with increasing exposure from 0.22 µg/l at 48 h to 0.055 at 96 h.

The results of continuous exposure showed that *Gammarus, Pteronarcys* and *Brachycentrus* were all sensitive to low concentrations, but each species responded differently to different exposure lengths. One conclusion from these experiments was that the LC_{50} values in the continuous exposure tests do not completely describe the toxic effect concentration for this insecticide. For example, in the case of *Pteronarcys*, 96 h exposure to 0.59 µg/l did not result in death; but the addition of 3 more days of constant exposure resulted in a 7-day LC_{50} of 0.071 µg/l, which is about 8 times less than the 4-day no-death concentration of 0.59 µg/l.

Continuous exposure produced an LC_{50} of 0.015 µg/l by the 18th day of exposure. This value is about 39 times lower than the no-death, 4-day exposure value. In the continuous exposures, no deaths were recorded at the highest concentration, 0.59 µg/l, at the end of 4 days, so that concentration was non-lethal. But when the test animals were transferred to clean water, they began to die, 70% by the 5th day, and 80% by the 6th day, demonstrating that 0.59 µg/l is in fact a lethal concentration.

Additional tests in this series, with reference to the synthetic pyrethroids permethrin and fenvalerate, have further endorsed the importance of exposure period in assessing behavioural and toxic effects (Anderson, 1982). For example, in the case of fenvalerate, three levels of sensitivity were observed. The most sensitive organisms were the amphipods, *Gammarus pseudolimnaeus*, in which toxic effects were observed within hours of exposure to low concentrations. Within 4–7 days over 50% of the gammarids had died after exposure to a concentration of 0.022 µg/l.

The second level of sensitivity was shown by the aquatic insects tested, all of which required a longer exposure time before death occurred. The LC_{50} values for the mayflies and *Atherix* (Rhagionidae) had reached 0.16 and

0.12 µg/l respectively in 7 days. By that time well over 50% of the amphipods were dead after exposures to 0.022 µg/l. By day 14, 80% of the mayflies were dead at that concentration, 0.022 µg/l. In contrast, 28 days of exposure were necessary to produce an LC_{50} of 0.029 µg/l for *Atherix*.

The third sensitivity level was recorded in snails – *Heliosoma* – which remained resistant to fenvalerate at 0.79 µg/l during the 28 days exposure.

In the case of permethrin, the response to low concentrations (0.029–0.52 µg/l) differed from that of fenvalerate at similar concentrations, with death not occurring until late in the exposure. Lethal concentrations (LC_{50}) could not be calculated until day 21 exposure in the case of *Brachycentrus*, and could not be calculated at all for the *Pteronarcys* exposures. However, in both species behaviour responses and signs of distress were evident long before death occurred. Those responses, particularly those inhibiting feeding, could well have contributed to the ultimate mortality in the last week of exposure.

In contrast with the comparative scarcity of data for flowing water systems, long exposures of 24, 48 and 72 have long been accepted practice in the technically less demanding standard static water tests. Although the relevance of static water test data to running water invertebrates is debatable, there are useful lessons to be learnt from recent reappraisal of the standard static water test *per se*. The most widely accepted practice in those procedures is to expose organisms continuously to a range of dilutions of the toxic chemical until they die, cumulative mortality being noted in relation to time.

A recent modification of this has been developed in which the test animals are exposed briefly to the toxicant and then transferred immediately to clean water (Abel, 1980). The object of this was to bring the test more into line with field reality where running waters are subject to occasional or accidental short-term contamination followed by relatively long periods of freedom from toxicity. The test organism selected was *Gammarus pulex* which played so useful a part in the flowing-water tests described earlier in this section. The effects of lindane (hexachlorocyclohexane) was tested firstly by the conventional continuous exposure method, in which continual observation enabled individual times of death to be determined, and secondly by the technique of brief exposure followed by a holding period in clean water. The results of the first conventional test agreed closely with those previously reported on toxicity of lindane to *Gammarus* spp (Cope, 1966; Sanders, 1969; Macek et al., 1976; Bluzat & Senge, 1979) and do not call for any further discussion in the present context.

In the second or modified test the animals were exposed to each of 5 concentrations from 2.0 to 0.1 mg/l for each of 6 time periods 1, 2, 5, 10, 20 and 50 minutes). Additionally, animals at the 3 lowest concentrations were exposed for periods of 2 h and 4 h, and up to $16\frac{1}{2}$ h (1000 min) in a

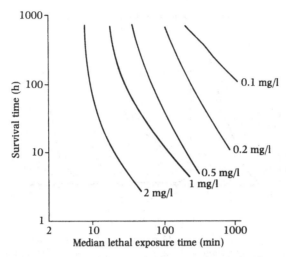

Fig. 5.2. Log-log plot of curves showing relationship between median lethal exposure time and survival time for *Gammarus pulex* exposed to five concentrations of lindane (2 mg/l to 0.1 mg/l); (after Abel, 1980).

further test. The observation period continued as long as control mortalities remained at 0 (14 days in one experiment and 21 days in another). The results (Fig. 5.2) showed consistently that at the higher concentrations of lindane some animals died during the exposure period. In the lower concentrations generally no animals died until several days in clean water had elapsed. Mortality among the organisms exposed to lindane continued to occur over the complete duration of the observation period, i.e. up to 21 days after exposure. The data in this series were used to calculate the values of a 20-day median lethal exposure time, i.e. the exposure time required to kill 50% of the animals within 20 days following exposure.

Evidence from other sources indicates that such relations are not specific to lindane, or to chlorinated hydrocarbon insecticides; studies on the impact of an organophosphorus insecticide, fenitrothion, on select aquatic invertebrates in Canada (Flannagan, 1973) point to the same conclusion. This is particularly well marked in the case of *Daphnia pulex*, *Lymnaea elodes* and *Gammarus lacustris* (the other test organisms being mosquito larvae). In this series the animals were exposed to various concentrations of fenitrothion for 24 h, removed from the test chamber and observed in clean water in the laboratory for a further 72 h. The results are shown in Fig. 5.3 and are of special interest in the case of *Gammarus lacustris*, more directly comparable to the data for *Gammarus pulex* already discussed. The figures show that over the range of fenitrothion concentrations of 6.0–50 ppb, 0 mortality was recorded after 24 h exposure. But the 96-hours cumulative mortality – 24 h

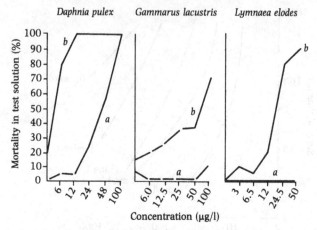

Daphnia pulex   *Gammarus lacustris*   *Lymnaea elodes*

Fig. 5.3. Curves indicating 24 h and 96 h cumulative mortality of *Daphnia pulex*, *Gammarus lacustris* and *Lymnaea elodes* exposed to various concentrations of fenitrothion for 24 h, then returned to clean water conditions. In each panel: *a*, mortality after 24 h exposure; *b*, mortality of same animals after 96 h, i.e. 24 h in fenitrothion solution +72 h in water (after Flannagan, 1973).

in fenitrothion + 72 h in clean water – revealed a delayed effect, with mortalities ranging from approximately 20 to 35%. With further increases in concentration to 100 ppb, the 24-h mortality of 10% increased to 70% after a further 72 h.

In the case of *Lymnaea elodes*, 0 mortality to a range of concentrations from 3.0 to 50 ppm after 24 h exposure, contrasts with the increased mortality revealed after a further 72 h, reaching a maximum of 90% in the case of organisms previously exposed to the highest concentration, 50 ppm.

Gammarus species are among the more robust and adaptable of stream organisms to laboratory conditions, both static and flowing. While most other stream invertebrates would be less suited to this type of long-term test because of increasing mortality in controls after about 1 week, there are others – such as dragonfly naiads – which would be admirable subjects for further study. One question which must be asked is whether these significant findings based on the long post-treatment holding period in clean static water, can be directly translated to events in the natural running water habitat. The first simple step towards clarifying this point would be the repetition of such tests under conditions in which test animals were exposed to continuous flow in both exposure and holding periods. Until that is done, the real significance of these findings cannot be fully assessed. What is certain, however, is that this study has revealed one more factor which cannot be ignored in any future evaluation of pesticide impact.

INVERTEBRATE DRIFT PATTERNS

BACKGROUND

The increase in invertebrate drift or benthic drift following application of pesticide to running waters has been recognised since the early days of DDT usage. In some of the earlier classical projects aimed at controlling *Simulium* larvae, such as those in Guatemala (Fairchild & Barreda, 1945) and East Africa (Garnham & McMahon, 1946) the massive DDT application rates – up to 20 or 30 ppm on occasions – produced a total catastrophic drift on such a scale as to be easily visible. In recent years, application rates, whether direct or indirect, have been pitched at much lower levels. Nevertheless, pesticide-induced drift remains a characteristic feature of invertebrate reaction to a wide range of toxic chemicals. The sampling of these drift organisms in relation to degree and timing of pesticide application is one of the main field evaluation techniques in streams and rivers. As a result, there is now an impressive amount of information from different study areas in many countries about the pattern of drift and its composition, according to the pesticide involved and its dosage regimen. That information on field studies is best described and assessed against the background of each project, and is reviewed in depth later (Part III).

Natural invertebrate drift (as distinct from pesticide-induced) has long been a subject of great interest to freshwater biologists. Here too there is now an impressive record of information about natural drift patterns of different organisms, the extent of regular nocturnal increase, and the role of endogenous and exogenous rhythms in determining the pattern. This information about natural drift patterns has been increasingly recognised and utilised by pesticide ecologists, who are fully aware that the drift patterns observed in contaminated streams are superimposed on a natural drift rhythm. In some evaluation programmes drift sampling has actually been concentrated on that period after sundown when there is normally a sharp increase in natural drift. In this way large numbers of organisms can be caught in a comparatively short trapping period, providing adequate data for valid comparisons between treated and untreated streams, or between pre- and post-treatment in the same stream or river.

ANALYSIS OF LABORATORY DRIFT REACTIONS
OF *SIMULIUM* LARVAE

The role of controlled laboratory experiments in defining drift patterns in relation to precise pesticide concentration was first recognised in practical studies on *Simulium* larvicides. *Simulium* larvae were established in artificial channels or troughs, initially alongside natural streams (Travis & Wilton 1965; Wilton and Travis, 1965) but laterally in the laboratory itself (Jamnback & Frempong-Boadu, 1966) permitting controlled dosage of larvicide to flow through the channel for specified periods.

These laboratory tests provided the first demonstration that at *Simulium* larvicide dosage, usually in the range 0.05–0.1 ppm, some insecticides had a more irritant or activating effect than others, as manifested in intensified downstream drift. Although it was noted that a few individuals which detached after exposure to the insecticide managed to survive, the general conclusion was that with organochlorine, organophosphorus and carbamate larvicides tested, detachment and drift were equivalent to mortality and could be used as a direct measure of lethal effect.

The development of an improved technique whereby large numbers of *Simulium* larvae were induced – without any handling – to attach themselves along the length of a miniature simulated stream 30 cm long and 3.8 cm wide, provided one essential requirement for further critical analysis of drift, namely the maintenance of control mortalities at or near 0 for the 3–4 days duration of each test (Muirhead-Thomson, 1969, 1973). The small size of the experimental channels (see page 42) also enabled a larger number of test units to be set up in limited laboratory space, and made it possible for a uniform batch of larvae from field sources to be distributed over several units and exposed to different concentrations simultaneously, under strictly comparable conditions. Even more important in the context of this review was that experiments with *Simulium* larvae led later to an extended programme on drift patterns in other 'non-target' macroinvertebrates under controlled laboratory conditions. Many of the observations in the early series have already been reviewed (Muirhead-Thomson, 1971), but the most instructive for the present purpose is a comparison between the reactions of *Simulium* larvae to two main groups of chemicals used most extensively in Simulium control, namely the organochlorine methoxychlor and the organophosphorus insecticides, including Abate(temephos). Those particular results justify tabulating in some detail (Table 5.5) so that they can be compared with the extensive drift net data now available from field trials.

The first point to note is that separate records of drift and mortality were made for both late instars ('full-grown') and early instars; this need being dictated by the established fact that early instars are relatively much more

Table 5.5. *Drift patterns, and subsequent 24-h mortalities, of* Simulium *larvae (mainly* S. ornatum) *exposed to methoxychlor (25% EC) and Abate (20% EC) in a laboratory-simulated stream*

Methoxychlor concentration (ppm)	Exposure period (min)	Number of larvae (total)	Cumulative percentage of larvae drifting after initial exposure to larvicide					24-h mortality (%)	
			0–15 min	0–30 min	0–45 min	0–60 min	0–2 h	0.2 h	Total
0.05	60	237	1	8	29	40	65	5	5
0.10	60	1239	6	19	51	83	92	20	
0.2	30	1148	17	62	84	87	90	59	
0.2	30	427[b]	39	63	73	75	78	80	

Control (number of larvae in total)	Cumulative percentage of larvae drifting			Total 24-h mortality (%)
	0–1 h	0–2 h	0–24 h	
1216	0.3%	0.3%	6.5%	0.25%

Abate concentration (ppm)	Exposure period (min)	Cumulative percentage of larvae drifting after initial exposure to larvicide			Total mortality 24-h (%)
		0–30 min	0–60 min	0–2 h	
0.2	60	0	0	0	90
0.2	60[c]	0	2	22	100
0.5	60	0	7	90	100
0.5	60[c]	17	35	97	100
1.0	30	0	23	51	99
1.0	30[c]	27	35	94	100

[a] After Muirhead-Thomson, 1977, and unpublished. (Figures refer to late instar larvae unless otherwise stated.)
[b,c] Early instars.

susceptible to most larvicides. The figures confirm and extend the known irritant effect of methoxychlor over the range of concentrations tested. At the higher dosage, 0.2 ppm signs of irritation became evident within a few minutes of introduction of larvicide dilution into the flowing-water system, and the first detachment occurred with 5–6 min. By the end of 1 h, detachment rates had reached a level of 80–90% in all tests.

The figures show clearly that despite high irritability and detachment produced, the drift rate is not automatically a measure of acute toxicity. With methoxychlor at the rate of 0.1 ppm for 1 h for example, over 90% of the late instar larvae had become detached by the end of 2 h, but the subsequent fate of those drifters – which had been removed to clean water – showed that 80% of those ultimately survived for 24 h at least.

In the case of the more sensitive early instars, it is seen that despite the very high detachment rate, mortality was never complete, 20% of this group surviving the 24-h holding period. The same general pattern of high irritability and detachment followed by significant survival was also exhibited in a short series of tests with the allied chlorinated hydrocarbons DDT and Gammexane (lindane).

In contrast, the results produced with several organophosphorus compounds, including Abate, revealed a very different pattern in which very little detachment took place up to 30 min after introduction of the larvicide dilution. After that time there was increasing drift on the part of early instars followed later by increased drift of late instars. The subsequent fate of drifters shows however that the majority of them had already absorbed a lethal dose of the chemical resulting in high mortality rates by the end of the 24 h holding period. This is most clearly marked at the two lower dosages used, 0.01 and 0.02 ppm in which 96–8% of late-instar larvae which had drifted within the first 2 h subsequently died, whereas the overall mortality rate of all larvae originally exposed was much lower (53–67%). At those concentrations the high mortality in drifters does not necessarily indicate an equally lethal effect on the total larval population at risk, this only begins to occur at the higher dosage rate of 0.05 ppm.

DRIFT REACTIONS OF MACROINVERTEBRATES
TO TEMEPHOS AND CHLORPYRIFOS

In a further series of laboratory studies in invertebrate drift (Muirhead-Thomson, 1978*a,b*) a slight modification in the testing procedure (see page 43) allowed simultaneous experiments to be carried out on several 'non-targets', in particular the amphipod *Gammarus pulex*, the ephemeropteran *Baetis rhodani*, the web-spinning caddis *Hydropsyche pellucidula* and the case-bearing caddis *Brachycentrus subnubilis*. The procedure was to run tests with *Simulium* larvae in a miniature simulated stream as described above, and parallel tests on other macroinvertebrates in the modified stream.

The first series of tests were aimed at comparing the reactions of test organisms to temephos on the one hand – the main larvicide in practical *Simulium* control – and another organophosphorus compound chlorpyrifos a candidate larvicide. More in keeping with field practice, the exposure period

Table 5.6. *Drift patterns of selected stream macroinvertebrates during a 30-min exposure to the organophosphorus larvicides chlorpyrifos methyl and temephos (Abate) in a laboratory-simulated stream, and subsequent 24-h mortality*

	Cumulative percentage drift at intervals after beginning of exposure		24-hour mortality (%)
	0–15 min	0–30 min	
Chlorpyrifos (0.1 ppm)			
Gammarus pulex	43	74	100
Gammarus (control)	0	1	0
Baetis rhodani	25	40	100
Baetis (control)	2	3	2
Simulium spp (late instars)	0	0	96
Brachycentrus subnubilis	0	0	0
Hydropsyche pellucidula	0	0	0
Chlorpyrifos (0.5 ppm)			
Brachycentrus	1	1	98
Temephos (0.1 ppm)			
Gammarus	49	70	0
Baetis	20	24	91
Simulium (late instar)	0	0	60
Temephos (1.0 ppm)			
Simulium (late instar)	0	0	99
Simulium (early instar)	0	27	100
Brachycentrus	0	0	100
Hydropsyche	0	0	62

was reduced to 30 min. The results with *Simulium equinum* (Table 5.6) confirm the same general reaction to organophosphorus compounds described above. With temephos there is no detachment of late instars in the first 30 min after exposures to concentrations, 0.5 and 1.0 ppm, which subsequently produced 100% mortalities, and the same pattern is shown by the lethal concentration of chlorpyrifos, 0.1 ppm. In the case of early instars, detachment in the first 30 min increased from 0 at 0.2 ppm to 17% and 27% at concentrations of 0.5 and 1.0 ppm.

With *Baetis* the drift response to both chemicals at the 0.1 ppm level, which produces a high mortality, is much the same as with early instar *Simulium* larvae, drift taking place in the first 15 min and increasing progressively thereafter up to 60 min. Extended tests with chlorpyrifos (Table 5.7) showed

Table 5.7. *Drift patterns of selected stream macroinvertebrates during and subsequent to a 30-min exposure to chlorpyrifos methyl in a laboratory-simulated stream and subsequent 24-h mortality*

	Cumulative percentage drift after introduction of larvicide			24-h mortality (%)
	0–30 min	0–60 min	0–24 h	
0.1 ppm				
Gammarus	69	75	77	100
Baetis	58	74	74	100
Simulium (late instars)	0	6	76	95
0.5 ppm				
Brachycentrus	0	0	0	97
Hydropsyche	0	0	0	64

however that even after exposure to this lethal dose there is very little increase in the peak drift achieved, 23% of *Baetis* still remaining in the channel.

A very different behaviour reaction was shown by the trichopterans *Hydropsyche* and *Brachycentrus* which showed almost complete absence of drift during 30 min exposures to concentrations of both chemicals which subsequently produced high mortality. In the case of chlorpyrifos, extended observation showed that drifting of both species remain at or near 0 right up to the end of the 24 h observation period following exposure to 0.5 ppm, the mortalities at this point being 64% and 97% respectively (Table 5.7).

Perhaps the most unpredictable behaviour reaction was that revealed by experiments with *Gammarus* when exposed to these two organophosphorus larvicides, bearing in mind its high sensitivity to one chlorpyrifos, and its high tolerance level to the other, temephos (see page 64). Exposure to both chemicals produced a very similar drift pattern (Table 5.6) irrespective of the fact that the concentration of chemical, 0.1 ppm, was one that produced 100% mortality in the case of chlorpyrifos and 0% in the case of temephos.

The extreme activation of *Gammarus* exposed to the insecticide stream also expresses itself in another direction, namely in vigorous escape reactions and attempts to leave the water. On the glass surfaces of the miniature simulated stream these attempts were unsuccessful, but where there is any foothold such as the gauze netting of the funnel trap (see Fig. 3.3) many *Gammarus* managed to crawl up the damp netting and succeed in escaping from the water. These reactions were just as marked on exposure to the lethal concentration of chlorpyrifos as to the completely non-lethal concentration of temephos.

These half-crawling, half-swimming movements of *Gammarus* in the experimental channel also enable the animal to move upstream. The natural upstream movement of *Gammarus* in its running water habitat has been well known to freshwater biologists for some years (Hultin, 1968; Hughes 1970; Elliott 1971 *b*; Wallace, Hynes & Kaushik, 1975) as has also been the natural upstream migration of other macroinvertebrates such as mayfly and stonefly nymphs.

This particular activity of *Gammarus* has been used as a measure of response to chemical in a quite different type of laboratory experiment (Ruber & Kocor, 1976; Thayer & Ruber, 1976) which is of special interest in this context in that it too was concerned with temephos and chlorpyrifos – chlorpyrifos ethyl (Dursban, in this instance). The experimental stream took the form of a circular channel in which a current of 9–15 cm/s was maintained by a paddle-wheel device. The test organisms, *Gammarus fasciatus* in groups of 100 were placed at one end of a 120-cm straight section of the experimental stream, and measurements were made of their upstream movements. The concentrations of insecticide selected were at sublethal levels of LC_{10} (24 h) which previous toxicity tests had established to be 2.0 ppm for temephos and 0.01 ppm for chlorpyrifos (a very wide difference in susceptibility similar to that recorded above for *Gammarus pulex*). Both treatments were found to inhibit significantly upstream movement relative to control values, the same effect being produced by concentrations of chlorpyrifos at least 10 times lower than those of temephos. As higher migration rates normally took place at lower rates of flow it was found that, in general, the inhibition of upstream movement was less obvious at high flow rates than at lower.

LABORATORY DRIFT STUDIES BY FRESHWATER BIOLOGISTS

The value of laboratory experimental streams in analysing drift patterns of stream macroinvertebrates has also been fully realised by freshwater biologists in their studies on natural drift. Much of that work has been aimed at the phenomenon of nocturnal drift, the relation between activity patterns and diel periodicity in drift, and the extent to which endogenous and exogenous factors are involved (Elliot, 1967*a*,*b*, 1968*a*,*b*; Chaston, 1968; Wallace *et al.*, 1975). Although it is not possible within the compass of this book to discuss that work in depth, it is clear that those observations, quite unconnected with pesticide impact, have considerable relevance to the laboratory evaluation methods discussed in this section as well as to the interpretation of experimental data. Methods devised for continuous observation of organ-

isms in laboratory streams throughout the 24 h, and under different illumi-
nation conditions, could all be used to advantage. It is also clear from the
work of those freshwater biologists that the activity of such organisms as
Gammarus in experimental channels or 'stream tanks' can be greatly
influenced by such factors as current, food, light, substrate and time of year
(Wallace *et al.*, 1975; Thayer & Ruber, 1976).

In view of all these variables it appears that, once again, the laboratory
methods developed to study impact of pesticide under 'simulated-stream'
conditions must be regarded as exploratory and open to considerable improvement. Despite so many variables it has been, however, the experience of
freshwater biologists that there is a remarkably close accord between the
activity and drift patterns observed under laboratory conditions and the
natural drift patterns recorded in the field (Elliott, 1968*b*). These experiences
also provide ample endorsement of the guiding principle in this phase of
evaluation, namely, that invertebrates which normally live in streams and
rivers can only be realistically tested in flowing water systems and not in
static ones.

DRIFT REACTIONS TO SYNTHETIC PYRETHROIDS

The series of tests discussed in the last section with reference to temephos and
chlorpyrifos disclosed consistent differences in drift reaction on the part of
different macroinvertebrates. Further variations in drift patterns were
demonstrated in a series of experiments with the synthetic pyrethroid
permethrin (Muirhead-Thomson, 1978*a*). The high toxicity of this chemical
to mosquito larvae and several non-targets organisms had already been
demonstrated (Mulla, Darwazeh & Majori, 1975; Mulla & Darwazeh, 1976;
Mulla *et al.*, 1980) while in laboratory simulated streams it had been shown
to be highly lethal to *Simulium* larvae (Muirhead-Thomson, 1977), being
about ten times more toxic than chlorpyrifos and about 40 times more than
temephos.

On the basis of these susceptibility tests, two concentrations of permethrin
were selected for drift studies, namely 0.005 ppm (5.0 μg/l) and 0.0005 ppm
(0.5 μg/l), an exposure period of 30 min being followed by a 24 h observation
period as in previous standard tests (Fig. 5.4).

During the short 30-min exposure period *Gammarus*, *Baetis* and *Simulium*
all showed a considerable degree of activation leading to downstream drift.
Again, in the case of *Gammarus* (in which activity started within a few
minutes of exposure), there was no great difference in drift pattern following
the high dosage which eventually resulted in 100% mortality, and the lower
dosage which produced only 32% mortality, the drift rates at the end of the
30 min exposure period being 84–94%.

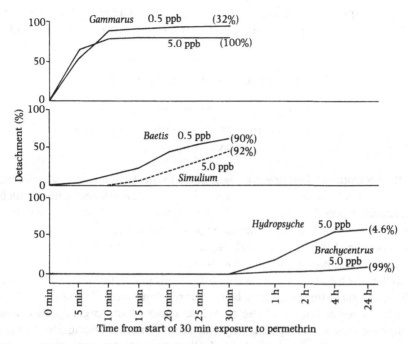

Fig. 5.4. Reactions of five stream macroinvertebrates in a laboratory-simulated stream, during and subsequent to 30 min exposure to permethrin at 5 ppb and 0.5 ppb. Figures in brackets are 24 h mortalities (after Muirhead-Thomson, 1978).

During this same period no drift at all occurred with *Hydropsyche* and *Brachycentrus*. However, when observations were continued for a further 24 h, during which they were once more exposed to a flow of clean insecticide-free running water, differences in drift pattern appeared (Fig. 5.4).

While *Brachycentrus* continued to show very little activation, with a drift of less than 10% after 24 h – i.e. on the part of larvae which had left their cases – *Hydropsyche* began to show signs of irritation shortly after the end of the 30-min exposure. As a result, many of the larvae left the shelter of the stony substrate and became caught up in the rapid flow of the current. Thirty minutes after clean water had been introduced at the end of the exposure period, significant drift had started reaching a maximum of 55%, 2–4 h after first contact with the insecticide. By the end of 24 h in clean flowing water, there was only a further slight increase in the drift. At this concentration of permethrin – 0.005 ppm – only 4.6% of the *Hydropsyche* had died after 24 h, compared to the almost total mortality in *Brachycentrus*.

This delayed activation and drift of *Hydropsyche* larvae, even to sublethal concentrations of permethrin, is in sharp contrast to its reactions to the organophosphorus insecticides temephos and chlorpyrifos, in both of which drift remained negligible up to 24 h at least after exposure to the insecticide.

VARIABLE BEHAVIOURAL REACTIONS

IMMOBILISATION, IRRITABILITY AND ESCAPE REACTIONS

In the previous section, reference was made to the extreme activity exhibited by *Gammarus* in particular when exposed to a larvicide stream, actions which inevitably exposed it to the flow of water, leading to drift: but the activities also took the form of vigorous attempts to leave the water by crawling, attempts which in a natural habitat could well prove successful on a mud or gravel substrate at the edge of a stream.

In the South African experience with dragonfly naiads (Muirhead-Thomson, 1973) several distinctly different behaviour reactions were observed according to the chemical nature of the pesticide. Three of the organophosphorus compounds tested – azinphos methyl (Gusathion), monocrotophos (Nuvacron) and Dipterex – caused the naiads of all species to show vigorous escape reactions which led them to crowd round the gauze-covered outlet to the flow-through chamber. This was not a manifestation of poisoning as there was a high survival rate to Nuvacron and Dipterex 24 h after a 1-h exposure to 1 ppm. No such reaction was produced by the other organophosphorus compounds tested, fenthion and parathion. Nor was there any sign of irritability on the part of organisms exposed to DDT such as was marked in the case of *Simulium* larvae. Of particular interest was the sharp behavioural difference of naiads when exposed to two different formulations of DDT. As far as mortality was concerned, there was no difference in effect between the two formulations at the high concentrations tested, However with a 50% emulsifiable concentrate formulation, concentrations of 10 and 20 ppm produced a progressive immobilising effect on the naiads during the 1-h exposure period, creating the impression that the insects were moribund. This condition continued after the end of the exposure period well into the holding period in clean water. However, a high proportion of the naiads eventually recovered completely. With a 75% wettable powder formulation, there was no indication of any phase of paralysis or immobilisation, and it can only be assumed therefore that one of the chemical additives in the EC formulation must have been responsible.

A similar effect was produced by the carbamate chemical tested, propoxur (Baygon) which, at concentrations of 5 ppm upwards, produced an immo-bilising or partly immobilising effect on the naiads, by the end of the exposure

Table 5.8. *Case-leaving responses and 24 h mortality of larvae of the caddis fly,* Brachycentrus subnubilis, *exposed to different time/concentrations of permethrin*

Concentration of permethrin (ppm) & period of exposure (min)	24-h mortality (%)	Dead out of cases (%)
0.001 (60)	64	60
0.005 (30)	100	35
0.0005 (30)	41	0

period. Here again, there was a high recovery rate by the end of the 24-h holding period in clean water, and most of the survivors were alive and active 4 days after the end of the experiment.

CASE-LEAVING BY CADDIS LARVAE, BURROW-LEAVING BY CHIRONOMIDS

A different type of behavioural reaction was observed in the case-bearing larvae of the caddis fly *Brachycentrus subnubilis* (R. C. Muirhead-Thomson, unpublished results). These are among the most active of case-bearing species, and are continually moving around on the aquatic plants, mainly *Ranunculus*, which provide an ideal substrate. With the different larvicides tested, a variable proportion of larvae were observed to leave their cases and eventually die. This was particularly marked with the pyrethroid permethrin (Table 5.8; R. C. Muirhead-Thomson, unpublished results).

From these figures there does not appear to be any consistent relationship between case-leaving and mortality. Although all larvae which leave their cases eventually die, case-leaving is not necessarily a measure of mortality among organisms exposed, as many larvae die without leaving their cases.

However, the case-leaving reactions of another species of *Brachycentrus*, namely the Canadian *B. numerosus* have been the subject of a more thorough investigation in New Brunswick (Symons & Metcalfe, 1978) with regard to the insecticide fenitrothion. Three series of experiments were carried out. First, test animals were exposed in a continuous test to different concentrations of fenitrothion in static (but aerated and agitated) conditions. Animals were examined twice a day for the exposure period of 5 days at the higher concentration, 15–20 ppm, and for 6 days at the lower concentration of 7.5–10.0 ppm, and a record kept of mortality and case-leaving.

In the second series, the exposure period – to a wider range of concentrations – was 24 h followed by a 72-h holding period in clean water. In the third series of tests, animals were maintained and tested on their natural rocky substrates in a laboratory stream tank, and recovery was recorded over a much longer period (namely, 4 weeks).

All larvae which left their cases were found to be moribund and unable to build new cases or to utilise empty ones when given the opportunity. In general, the case-leaving response increased with increasing concentration, but the percentage leaving their cases varied in different experiments at the same concentration.

Other species of caddis fly larvae respond to stress by leaving their cases, as reported for *Brachycentrus americanus* for example when exposed to endrin (Anderson & DeFoe, 1980), and serious consideration has been given to the possibility that this type of response, or another equally well-defined reaction, might provide a better guide to toxic exposure levels than mortality (Symons, 1977a,b). The first step in the development and use of such indicator organisms is to determine the EC_{50} – i.e. effective concentration for 50% of the organisms – for the case-leaving response of several species of caddis larvae, particularly those available during the period of a spray programme. In addition to *Brachycentrus numerosus* and its response to fenitrothion, another abundant species *Pycnopsyche gottifer* has been used to establish such levels for trichlorfon (Dylox). This case-leaving response is by no means peculiar to insecticides: about 50% of *Brachycentrus americanus* have been reported abandoning their cases and dying when exposed to the piscicide antimycin (Jacobi & Degan, 1977). Caddis larvae are similarly affected by the herbicide dichlobenil (Wilson & Bond, 1969). The ideal would be to find a series of species which leave their cases at different levels of exposure to pesticide, and then to calibrate the case-leaving response to recorded kills of free-living insects under controlled spray conditions in a number of streams (Symons, 1977b). Probably the greatest advantage of using such well-defined behaviour reactions would be when, and if, they occurred at sublethal levels and could be used to detect the presence of toxic chemicals at such low concentrations as are liable to be overlooked or undetected by routine physicochemical assay.

A behavioural response in somewhat the same category is shown by the tendency of some chironomid larvae to leave their burrows when exposed to insecticide. This reaction has been examined, with particular reference to *Chironomus tentans*, as part of a laboratory investigation to compare the effects of two methoxychlor formulations on target *Simulium* and on non-target invertebrates in Canada (Sebastien & Lockhart, 1981). Larvae were allowed to establish themselves in silica sand provided as a substrate in laboratory beakers, after which methoxychlor at the rate of 0.1 mg/l and 0.3 mg/l (i.e. ppm) was added. In the case of the emulsifiable concentrate (EC) formulation, the dosages proved lethal – 98% mortality – by the end of a 96-h observation period; in the course of that period, there was a steady increase in larvae leaving their burrows and showing characteristic wriggling movements, which reached a peak of 98% after about 12–24 h. With a particulate formulation of the insecticide, a higher proportion of larvae left their burrows than eventually died, as follows, indicating that this behavioural response, in the case of the particulate formulation, was still operative at sublethal levels

Methoxychlor concentration (mg/l)	Larvae leaving burrows (%)	Mortality (%)
0.1	55	21
0.3	87	36

IMPLICATIONS OF BEHAVIOURAL RESPONSES

There is clearly a great need for more detailed knowledge about behavioural responses under pesticide stress. In the absence of such information one can only speculate as to the extent to which these responses protect the organisms or lead to further exposure to the chemical. For example it would seem reasonable *a priori* to conclude that organisms which are induced to drift within a short period of first exposure to toxic chemical, and in which considerable drift actually takes place during the passage of the insecticide wave, would continue to be exposed to the effects of the chemical for an extended period as they moved downstream along with the chemical. In contrast, species which were slow to activate and drift might enter the stream after the pesticide wave has already passed, and perhaps have an increased chance of survival. This might also apply to species which are simply immobilised by the presence of chemical, retaining their hold on the substrate.

This may be rather too simple an interpretation. Some of the species most ready to respond by detachment and drift, such as mayfly nymphs *Baetis* and the amphipod *Gammarus*, are also extremely active free swimmers; and it is easy to visualise how such active movements could take them out of the main pesticide stream into some protected refuge or backwater, or even – as in the case of *Gammarus* – enable them to leave the contaminated water temporarily.

BEHAVIOURAL REACTIONS AND PREDATOR/PREY
RELATIONSHIP

Many of the macroinvertebrates which constitute the normal drift pattern in streams show little tendency to expose themselves to the main stream of water during daylight hours. It would seem likely that any disruption in this pattern, such as that caused by inflow of toxic chemical, which causes the organisms to leave their safe daylight refuge would tend to expose them to predators to an increasing extent.

Some of the numerous field studies on changes in feeding patterns of fish on fish food organisms – as judged by examination of stomach contents – are described in Chapter 7, but there are perhaps rather fewer direct observations on this predator/prey situation under laboratory conditions. Two examples are particularly illuminating.

One of these is the effect of temephos (Abate) on the marsh fiddler crab (*Uca pugnax*) (Ward & Busch, 1976). Although this animal is not a freshwater organism, living as it does in salt marshes, its study is important because of the analytical laboratory approach, and because the chemical involved – Abate (temephos) widely used for control of salt marsh mosquitoes – is one whose impact on stream fauna in general has been extensively studied elsewhere. An important survival feature of this animal is its very rapid response to threatening moving objects which enables it to escape the danger of approaching predators, birds or humans, by quickly disappearing into the nearest burrow. This behaviour response had previously been shown to be impaired by ingesting sediments containing DDT (Odum, Woodwell & Wurster, 1969)

Groups of crabs, acclimatised for a week prior to testing, were exposed in glass test chambers containing marsh water to which various concentrations of temephos had been added. At the end of 24-h exposure crabs were removed, mortality recorded and survivors tested for escape response to a suddenly presented, rapidly moving specimen of a predatory bird, the willet. For this test, an enamelled pan lined with paper towelling moistened with marsh water, was arranged in a wooden box in such a way that the crab could not see any movement on the part of the experimenter.

Two lines, 30 cm apart, were drawn across; the response of the crab to the sudden appearance of the willet was measured by (a) whether or not the crab escaped across the far line within 15 s of presenting the 'bird' (b) by recording the time taken to cover the distance between the two lines. The baseline datum was provided by the fact that all healthy crabs crossed the far line within the 15 s period following stimulation.

The percentage mortality of *Uca pugnax* increased with increasing concentration of temephos (the LC_{20} for 24-h exposures was 2.06 ppm; the LC_{50} was

9–12 ppm and the LC_{80} was 39.8 ppm). In order to arrive at a realistic figure for *total* mortality it was necessary to add the percentage of crabs which did not respond to the stimulus by escaping – assuming that loss of escape behaviour would inevitably have lethal consequences in the field.

The results showed that when the direct mortality was 50% or less, the escape reaction is impaired and the animal is 'ecologically dead' in roughly twice as many crabs as would be expected from the mortality data alone. Field studies endorsed the fact that increased predation on *Uca* accounts for most of the mortality following treatment with temephos, and show that acute mortality data may seriously underestimate the impact of a pollutant on this species at least (Ward & Busch, 1976).

A second example from the laboratory is provided by the predation on chalk stream invertebrates in England by the bullhead fish (*Cottus gobio*). In laboratory cultures *Cottus* will feed readily on introduced larvae of the caddis *Hydropsyche* and *Rhyacophila*, but not the case-bearing *Brachycentrus* (Muirhead-Thomson, 1979), and also feeds readily on *Gammarus*. In these cultures *Gammarus* quickly finds shelter among stones or other substrates provided, or among the dense tangle of aquatic vegetation and roots of submerged plants. However, the addition of different toxic chemicals – particularly permethrin – leads to great activity and excitability, as observed in experimental channels. The *Gammarus* leave their shelters and become free-swimming in the open water of the aquarium where they are quickly attacked and devoured by *Cottus*.

REACTIONS OF SNAILS TO MOLLUSCICIDES

Aquatic snails are in general little affected by the wide range of insecticides in common use, but show different types of reaction to the particular pesticides, molluscicides, designed for their control. For many years, laboratory toxicity tests for aquatic snails have been based on static water containers under conditions in which there is a marked tendency for snails to crawl out of the test medium, making it necessary to return them to the test solution whenever this happens. It is only comparatively recently that this obvious defect in the static water test has been counteracted by the introduction of flow-through tests in which snails are confined to glass cells through which there is a continuous flow of pesticide dilution, enabling observations to be carried out without interruption over a period of several days (Duncan, Brown & Dunlop, 1977).

In the natural habitat where there is no such confinement, it seems likely that this escape reaction in presence of toxic chemical must, as in the case of the amphipod *Gammarus*, prompt some animals to leave the contaminated water medium of the stream, ditch or river, and perhaps survive for a period

91

on the moist stream bank before returning to the water when normality is resumed.

INDIVIDUAL RESPONSE IN ECOSYSTEMS

APPLICABILITY OF LABORATORY EVALUATION DATA

In Part III, evaluation methods and evaluation data from a wide range of field projects concerned with pesticide impact on running waters are examined in depth. In an ideal situation, these field studies should be assisted from the start by a solid foundation of information about the reactions of all members of the community to the toxic chemical under controlled laboratory conditions. Again ideally, this laboratory programme should involve exposure of individual organisms – under simulated-stream conditions – to the range of time/concentration exposures likely to be encountered in the natural habitat, as well as to the fluctuations, or progressively declining concentrations of the chemical following its initial injection. The scope of the laboratory evaluation would also allow for physical factors likely to affect the performance, persistence and distribution of injected chemical; with comparative tests at different temperature conditions, hardness and softness of the water, different water velocities etc., as well as studies on uptake, retention and loss of chemical from different organisms. However, these ideal requirements are rarely achieved. Perhaps only in the exceptional case of the evaluation programme dealing with the impact of the fish toxicants, TFM and TFM/Bayluscide, used for control of larval lampreys in the Great Lakes has a scientifically planned laboratory phase played a key role in the overall environmental assessment (National Research Council Canada, 1985). Some valuable lessons to be learnt from that project will be discussed.

As it is, the ideal is rarely feasible as it would involve a time and labour commitment which is not generally acceptable. In the majority of cases, the scientist concerned with evaluation in the field has only the assistance of fragmentary laboratory data – normally in the form of acute toxicity tests at standardised exposures of 24, and 48 h in static water. On occasions, even that basic information is lacking.

This anomalous state of affairs is well illustrated by the intensive field evaluation studies carried out in eastern Canada on the environmental impact on stream fauna of several insecticides used experimentally for control of spruce budworm (*Choristoneura fumiferana*, see Chapter 7). These investigations, spread over several years, have been outstanding for the quality of the sampling procedures and the interpretation of sampling data. However, as far as stream macroinvertebrates are concerned, there has been no corresponding laboratory phase of evaluation, and certainly little basic information on tolerance levels based on essential simulated-stream techniques. Such information would clearly have been extremely valuable in interpreting field sampling data and in assessing the ecological effect of low concentrations of pesticide in the water. In the absence of such baseline data, reference has had to be made to tolerance levels established on closely related species in another faunal region (in this case, England). The results obtained with three European species of *Simulium, Baetis* and the case-bearing caddis *Brachycentrus* were compared with observations made on the same genera – but different species – in the Canadian field trials involving the insecticide chlorpyrifos methyl (Reldan) (Holmes & Millikin, 1981) and permethrin (Wood, 1983).

In the case of chlorpyrifos methyl, laboratory simulated-stream studies on the European species had shown that nymphs of *Baetis* were readily activated by the insecticide leading to detachment and downstream drift, while activation and drift of *Simulium* appeared to be delayed (Muirhead-Thomson, 1978*b*). These observations were confirmed in the Canadian studies where peak drifts of baetids were recorded immediately after application, but not until 6 h later with *Simulium*. With regard to actual tolerance levels, the laboratory experiments had shown that *Baetis* was also more susceptible than *Simulium*; the Canadian observations however indicated that although drift reactions of *Simulium* were delayed and less marked, the response to treatment in the field was more marked with *Simulium* than with *Baetis*. The laboratory experiments had also shown that *Brachycentrus* showed little sign of activation by chlorpyrifos methyl, with the majority of exposed larvae dying *in situ*; the Canadian *Brachycentrus* were observed to drift in response to insecticide treatment.

Similarly, in a study of permethrin residues in New Brunswick streams (Wood, 1983) the likely ecological effect on stream invertebrates had to be gauged on the only available laboratory data, namely the LC_{90-s} figures for the European *Gammarus pulex, Brachycentrus subnubilis* and *Baetis rhodani* (Muirhead-Thomson, 1978).

Attention is drawn to these comparisons not so much in order to confirm or otherwise the relevance of laboratory data to field observations, but to emphasise once more how essential it is to have a laboratory phase as an integral part of any evaluation programme, and to plan this phase in close accordance with local stream fauna and their particular ecology and habitat requirements.

EVALUATION METHODS AND THE FORECASTING
OF EVENTS

Closely bound up with the need for a closer understanding of the relevance of laboratory data to natural conditions in the field, is the increasing emphasis, and increasing demand, in recent years for more precise methods of forecasting the course of events in contaminated running waters. In the case of target organisms, usually the subject of particular attention at field and laboratory level, it is possible in many cases to make a reasonable prediction as to the course of events following application of the appropriate pesticide to their aquatic habitats. This has been amply demonstrated for example by the continued investigations on control of the larvae of the sea lamprey (*Petromyzon marinus*) (National Research Council Canada, 1985). While it is possible under those conditions to predict the reactions of specific organisms, it is a very different problem to progress to the next logical step and interpret these findings in terms of community or ecosystem response.

The different organisms in the stream ecosystem do not all live in identical laboratory-simulated streams, but in a highly complex system in which their normal attachment sites and behaviour patterns introduce new and, in many cases, unpredictable elements into their response to pesticide impact. In addition, the time/concentration exposure to the contaminant chemical tends to vary widely in the cross section of the stream or river, and between surface and bottom layers, thus having a differential impact on species which are mainly restricted to those particular habitats.

EXPERIMENTAL APPROACH TO COMMUNITY
RESPONSE

It would seem that the first step in interpreting the results of single-species laboratory tests in terms of the complex ecosystem would be to establish communities of representative species in the laboratory in the form of a miniature ecosystem into which the toxic chemical can be introduced at controlled concentrations for controlled periods. This is the basis of the rapid-through-flow test chamber used to study the impact of *Simulium* larvicides on several representative non-target stream organisms (Muirhead-Thomson, 1973, 1978*b*). The principle has been taken further in a critical study on the effect of the insecticide diflubenzuron – a chitin-synthesis inhibitor (Hansen & Garton, 1982). In this, a series of standard single species toxicity tests were first carried out according to US standards, and then followed by exposing a wider range of species in continuous flow laboratory-

94

stream channels divided into riffle areas and pool areas. The wide and representative range of stream fauna established in this community channel included five species or specific groups of mayflies (Ephemeroptera), seven of stoneflies (Plecoptera), three of caddis (Trichoptera) as well as several Diptera and Coleoptera.

Of the wide range and representation of stream invertebrates in this community channel, only very few had first been toxicity tested by standard methods, LC_{50} values being obtained for only two insect (midge) and two crustacean (*Hyalella* and *Daphnia*) species. This made it difficult to throw light on the main problem regarding the relevance of standard toxicity data for single species to their ultimate response as members of a community. Nevertheless, the data from the community response *per se* to continuous exposure to four different concentrations of diflubenzuron, viz., 0.1, 1.0, 10.0 and 50 µg/l, represents a major breakthrough in this bridging phase of evaluation.

In order to compare the reactions of different organisms in terms of their place in the community, identified invertebrates from this continuous exposure experiment were placed in appropriate groups as (*a*) scrapers – a diverse group; (*b*) shredders – mainly some species of mayfly and stonefly; (*c*) predators – mainly stonefly, Diptera and beetles and (*d*) collectors – including gatherers and filterers, mainly some species of mayfly and stonefly. Responses according to these categories ranged from rapid but less-persistent effect in the first group, generally rapid and persistent effect in the second group, slow but continuous effect in the third group, and rapid but discontinuous effects in the fourth group. The gradual nature of the effects in group (*c*) was mainly determined by differences in sensitivity within the predominant stonefly component. Some stoneflies were rapidly eliminated, e.g. *Pteronarcys*, *Nemoura* and *Isoperla*, while others (such as *Acroneuria*) required several months of treatment before elimination occurred.

It was concluded from these experiments that tests with single species could predict effects due to direct lethal impact, as proved to be the case with the expected reduction in biomass and diversity of invertebrates known to be sensitive to diflubenzuron. However, these tests could not predict the exact nature of the effects produced in the stream community, nor the effects resulting from interspecies reactions. This unpredictable element is well illustrated by the fresh water algae whose inclusion is specified in US guidelines. Algae, as represented by *Selenastrum*, were not affected by diflubenzuron in the single-species tests, but in the community-stream stage an algal bloom (involving all the algal species) was produced, plus the appearance after two months of such algae as *Melosira* and *Synedra* which did not appear in controls for two months later. This increased algal bloom – which has been noted in other similar trials was attributed to the decreased browsing by the

95

susceptible invertebrate herbivores, leading in turn to increased competition for the limited attachment sites among the algal species, with subsequent increase in the successful competitors.

Another instructive example of the analysis of secondary effect of pesticide contamination of streams is provided by the studies on carbaryl (Sevin) in Maine, US (Trial, 1982*a,b*, 1984), one of several insecticides used in control of spruce budworm. It had already been established that one of the clearly defined effects of carbaryl impact was a reduction of stonefly (Plecoptera) populations, primarily by *Leuctra* and *Nemoura* (Trial & Gibbs, 1978; Courtemanch & Gibbs, 1978) populations remaining depressed for several years (Trial, 1978, 1979, 1980). Most of these species come into the category of 'shredders', feeding mainly on leaf litter in the stream. Studies on the ecology of natural streams, particularly on the trophic relations of aquatic insects (Cummins, 1973, 1974; Cummins & Klug, 1979; Wallace & Merritt, 1980; Cummins et al, 1983) indicated that the removal of leaf shredders from a stream will increase the amount of nutrients and coarse particulate matter favourable for grazers, and decrease the amount of algae and fine particulate matter favoured by 'collectors', i.e. those species which consume particles less than 16 mm after filtering them from the water or gathering them from sediment surfaces. Indications from these many different sources all pointed to the conclusion that shredders, including *Pteronarcys* significantly increased the nutrients available to collectors, the 'collectors' being further subdivided into 'filter feeders' (e.g. *Simulium* larvae) and 'gatherers'.

In the trials in Maine, it was found that, in undisturbed streams, the ratio of shredders to collectors was constant from year to year. In the year following carbaryl contamination, when populations of shredders were depressed, the predicted secondary effects did not materialise. Collector densities in sprayed streams remained equal to or greater than those in control streams. It was concluded that, in this instance, the predicted fall in numbers of collectors as a consequence of the decline in numbers of shredders did not take place because there was enough food available – produced by shredders in previous years – as well as food from sources other than shredding.

In the Onchocerciasis Control Programme (OCP) involving regular applications of *Simulium* larvicides – mainly temephos (Abate) – to extensive river systems in West Africa (Opong-Mensah, 1984), there was no true laboratory phase of evaluation and consequently no baseline data sufficient to establish LC_{50} values for invertebrates at risk. However, the systematic use of experimental channels installed in the river bed provided adequate information about the reactions of a wide range of invertebrate fauna to the dosage rates used in the practical control of *Simulium* larvae, viz., 0.1 and 0.05 ppm for 10 min, according to season.

Despite the many indications from these experiments about the immediate lethal effects of treatments on 'non-targets', sufficient to arouse misgivings about serious long-term environmental effects, recovery of the running water communities to normality took place within a matter of weeks following suspension of treatment. Practically the only evidence of possible ecosystem effect was provided by the fact that there was an actual increase in the numbers of some groups (such as tanytarsine midges) and that the elimination of the target larvae of *Simulium damnosum* led to an increase in some non-target *Simulium* species which must have moved in from microhabitats located in parts of the river less affected by treatments, and been able to occupy the niche left by the disappearance of the target species.

With regard to single-species reaction to pesticide exposure in the laboratory *vis-à-vis* their reactions to the same chemical in the field, generally this can be predicted fairly accurately with species which are highly sensitive to the chemical in question. In fact, this is the essential basis of chemical control of undesirable aquatic fauna whether they be insect larvae (mosquitoes, *Simulium*, nuisance midges and caddis), intermediate snail hosts or undesirable predatory or trash fish. There is now good reason to believe that the same direct relevance of laboratory data to field reality is equally well marked in the case of organisms which show an unusually *high* tolerance to a particular chemical in laboratory tests. This is well brought out in studies on the lamprey larvicide, TFM, in tributaries of the Great Lakes (National Research Council Canada, 1985). Laboratory investigations involving a good representative range of stream invertebrates (Maki, Geissel & Johnson, 1975) had shown a high tolerance level on the part of the amphipod *Gammarus pseudolimnaeus* (with a 24-h LC_{50} of 38.0 mg/l). Nevertheless, field reports continued to indicate that this species was killed by TFM treatment of streams, which are pitched at a much lower level of concentration, usually less than 10 ppm for 8–20 h. This anomaly was solved only comparatively recently (Maki, 1980, Kolton *et al.*, 1985) in which drift net studies of treated streams showed that the rapid fall in *Gammarus* density following treatment was mainly caused by the high excitability of this organism to the chemical. Previously, the high numbers of these detached *Gammarus* had been interpreted as proof of lethal effect, whereas in fact they indicated only a shift of hyperactive populations downstream.

Another example of the value – at least retrospectively – of laboratory-susceptibility tests is provided by some of the earlier reports, from Kenya, East Africa, on the use of DDT for control of *Simulium* larvae in rivers and streams (Garnham & McMahon, 1946; McMahon, Highton and Goiny, 1958). At that time excessively high dosages of DDT were used, usually around 5–6 ppm for 30 min, but increasing on occasions to levels of 20, 27 and 36 ppm in two main rivers. These applications produced high fish kills and what appeared to be an almost complete biocidal effect on all running water fauna. However, after six months' treatment, dragonflies were still observed flying over the controlled rivers. It was not until 30 years later that the opportunity arose to establish susceptibility levels of African dragonfly nymphs – both agrionid and libellulid – to various insecticides under controlled laboratory simulated stream conditions (R. C. Muirhead-Thomson, unpublished results). These tests showed that the nymphs of the four African species observed had consistently high tolerance levels to DDT, with high survival rates 24 and 48 h after a 1 h exposure to concentrations as high as 20 ppm, and almost total survival after similar exposures to 10 ppm, the final criterion of survival being the successful emergence of adults from nymphs in the laboratory.

LABORATORY SUSCEPTIBILITY AND COMMUNITY RESPONSE IN STATIC WATERS

Studies on carbaryl (Sevin) in Maine The problems of pesticide impact on static water bodies such as ponds, swamps, marshes and lakes have had to be excluded from the present review as limitations of space would make it impossible to do justice to the great volume of published work in that field. However, in the context of the present chapter, there are valuable lessons to be learned from some recent work on the relevance of laboratory-susceptibility test data to specific and community response. Although the macroinvertebrate fauna studied were characteristic of static water bodies rather than of running water, many of the species involved are closely related to typical stream representatives and may, on occasions, occur in the stagnant back-waters of rivers or in seasonal relict pools in river beds. One of the illuminating examples of the work carried out as part of a general aquatic programme in Maine was quoted earlier in this section with reference to stream invertebrates. For several years, continuous studies were also carried out on the effects of those chemicals on the fauna of small woodland pools, both sphagnum ponds with deep organic substrate and beaver ponds with predominantly mineral substrates (Gibbs *et al*, 1981; Gibbs, Mingo & Courtemanch, 1982, 1983, 1984).

Laboratory susceptibility tests on the main insecticide involved, carbaryl (Sevin), had shown that it was toxic to most pond invertebrates (Bluzat & Senge, 1979; Johnson & Finley, 1980). Preliminary observations on ponds in Maine had confirmed the lethal effect of Sevin contamination on immature mayflies and caddis flies, and particularly on the freshwater amphipod *Hyalella*. Experimental ponds were treated at rates commensurate with spray dosages in spruce budworm control and observed for a prolonged period of $3\frac{1}{2}$ years in order to establish immediate and long-term effects on the fauna. Over the same period, studies were carried out on the occurrence and persistence of carbaryl residues in pond water and in sediments, water samples being collected regularly both at the air-water interface and at a depth of 0.5 m.

As predictable from the laboratory-susceptibility tests, there was an immediate reduction in abundance of both immature and emerging Ephemeroptera and Trichoptera following treatment, but conditions returned to normal within a few weeks with no evidence of any long-term effects. In the case of the amphipods *Hyalella azteca* and *Crangonyx richmondensis*, the effects of treatment were catastrophic and persistent, and dead or distressed organisms were seen within a few hours of treatment. In one of the treated ponds, the numbers of *Hyalella* remained at or near zero throughout the rest of the treatment year (1980) and through 1981 and 1982, with no increases indicative of recolonisation until 1983. In the same treated pond, the numbers of *Crangonyx* diminished to zero in the first sample after treatment, and remained near zero from 1980 through, to and including 1983. In the second treated pond, which contained no *Crangonyx* originally, numbers of *Hyalella* also remained near zero until late 1981, with recolonisation in 1982.

Correlated with these observations was the unexpected long-lasting presence of carbaryl residues in the treated ponds, still detectable more than 426 days after treatment in the pond which initially recorded a higher carbaryl concentration. During that period the decrease of carbaryl in the water from the original concentration of 254 µg/l followed a steady course correlated with time. These original concentrations were greatly in excess of the 96-h LC_{50} values recorded for amphipods, viz., 22 µg/l and account in part for the devastating effect of carbaryl treatment. It is difficult to say whether the long-term effects produced were due mainly to the undetermined effect of low, persistent concentrations of the chemical, to other factors such as intense predation or reproductive failure at low densities, or to difficulties encountered by a purely aquatic organism (with no aerial phase) in recolonising the ponds. The persistence of residues in the water and not just in the sediments again emphasises the wide gap which can exist between impact of the same chemical on still and on running water habitats.

Studies on cypermethrin in the UK   The second example of pesticide impact on the macroinvertebrates of static water embodies all the features essential for the progressive evaluation from laboratory to field and, as such, contains many valuable pointers in the approach to allied problems in running water. The pesticide in question was the synthetic pyrethroid cypermethrin, used widely in controlling a variety of insect pests. In the first phase of evaluation (Stephenson, 1982) acute toxicity tests, using 24-h static procedures, were carried out on 10 freshwater invertebrates including the isopod *Asellus*, the amphipod *Gammarus*, and nymphs of the mayfly *Cloeon dipterum*. The toxic effects were monitored, namely reduced mobility and death. The actual concentration of cypermethrin was chemically assayed at the beginning and at the end of the 24-h test period, and confirmed an expected loss of toxicant likely to take place in such static tests; up to 30% at concentrations of more

Table 5.9. *Concentration of cypermethrin in surface and subsurface waters of ponds at various intervals after application of chemical*

Time after treatment	Concentrations of cypermethrin (μg/l)	
	Surface water	Subsurface water
1 h		0.8
4 h	100	2.3
24 h	13	2.4
48 h	4.8	2.6
6 days	1.4	1.0
13 days	0.7	0.9

After Crossland (1982).

than 1 μg/l, but even higher losses, 70–80% at lower test concentrations of 0.05 μg/l.

The LC$_{50}$ values obtained ranged from 0.1 μg/l for *Gammarus*, 0.2 μg/l for *Asellus*, 0.6 μg/l for *Cloeon* and approximately 5 μg/l for the aquatic beetle *Gyrinus*.

In the second phase of evaluation (Crossland, 1982; Shires 1983) three natural ponds (one 20×5 m and two at 10×5 m) were treated with cypermethrin, and observations made on the natural fauna and on the chemical residues. In connection with the latter, an important feature was that water samples for residues were taken at two levels, viz., surface 2.5–10 mm in depth, and subsurface at about 50 cm, a feature which was found to have an important bearing on subsequent interpretation of invertebrate reactions. The differences in concentration of cypermethrin at these two levels were as follows (Table 5.9).

As predictable from the results of the static water laboratory tests, the community of invertebrates in the ponds was markedly affected but there was considerable variation in the rate at which different insects and crustaceans were affected by the treatment. Aquatic insects which frequently came to the surface for air were affected most rapidly, such as late instar nymphs of the water boatman *Notonecta* and water beetles such as *Dytiscus marginalis* which were affected within 1 h of treatment and died at the surface of the pond. Dipterous larvae such as *Chrironomus* were not affected until 24 h after treatment.

This first experiment was terminated at the end of 2 weeks when the pond was pumped dry for removal and identification of fish. A second, longer-lasting pond experiment was carried out over a period of 16 weeks. In this experiment, the surface water sampled by chemical assay was taken from an

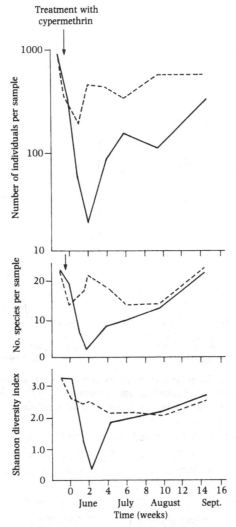

Fig. 5.5. Effect of cypermethrin treatment on macroinvertebrates in experimental ponds oversprayed at a dosage of 100
g/ha, showing rate of recovery (after Crossland, 1982). Broken
lines, untreated pond; solid lines, treated pond.

even narrower zone, i.e. to a depth of only 0.05 mm, and revealed even
higher concentrations than in the first test, as compared with the subsurface.
The effect on macroinvertebrate populations as a whole is shown in Fig. 5.5

The effect of the treatment on the zooplankton was, as in the case of insects
and crustaceans, in conformity with the results of laboratory-susceptibility
tests which had indicated that they would be unlikely to survive 48-h
exposures to concentrations exceeding 0.5 μg/l of cypermethrin. Daphnids

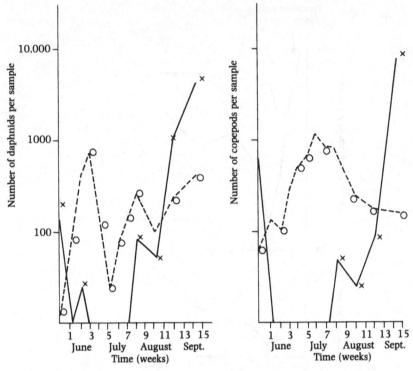

Fig. 5.6. Effects of similar treatment as in Fig. 5.5 on daphnids
and copepods in experimental ponds (Crossland, 1982). Broken
lines, untreated ponds; solid lines, treated ponds.

and copepods were not found in treated ponds until 8 weeks after treatment,
after which there was an increase in numbers far exceeding that in an
untreated pond (Fig. 5.6). Associated with this was the increase in mass of
filamentous algae in both of the treated ponds, possibly as a result of the
immediate effect of treatment on fauna such as the mayfly larvae *Chloeon
dipterum* which normally exert grazing pressure on the algae.

The residue studies also showed that the strong affinity of cypermethrin for
surfaces, particularly those – including suspended solids – with a high content
of organic matter, limited its downward dispersion into subsurface waters.
The overall effect was that the concentration of cypermethrin in subsurface
water, which showed a maximum of 2.6 µg/l 48 h after treatment, was
much less than that calculated from the known application rate, viz.
16–18 µg/l. Although this partial removal of cypermethrin from subsurface
water was not enough to upset the course of events predicted from the
laboratory tests on invertebrates, it was evidently sufficient to account for a
marked discrepancy between laboratory test results with a fish, the rudd
(*Scardinus erythrophthalmus*) which was found to survive concentrations of

102

cypermethrin in the ponds greater than the LC_{50} values determined in the laboratory with filtered mains water.

Those pond experiments, following the laboratory-testing phase, provided an invaluable step in a sequence of studies aimed at evaluating the hazards of cypermethrin as used in practical control of agricultural pests (Crossland *et al.*, 1982; Shires & Bennett, 1985). In those studies, which also extended to running waters (see page 67) it was found that in normal pest control spraying operations, involving application of cypermethrin by both tractor-mounted sprayer and by aerial spraying, very low residues were found in the subsurface water of static ponds, the maximum 5 h after treatment being of the order of 0.03–0.07 µg/l, with the lowest figures approaching the limits of detection, namely 0.01–0.02 µg/l.

Only minor and transient effects on aquatic macroinvertebrates were observed. In order to check how far minor population changes or behaviour reactions were attributable to the presence of cypermethrin at such low concentrations, an additional evaluation technique was introduced at this stage, namely bioassay using *Gammarus pulex* as indicator: the term 'bioassay' in this context being used in its strict sense, namely the use of a living indicator organism – whose susceptibility levels to the pesticide in question have already been established – as a biological indicator of chemical presence or concentration.

This is a particularly valuable technique at low toxic chemical concentrations, approaching the limits of chemical assay, but which may still be biologically significant. The 24-h LC_{50} of *Gammarus pulex* had been established as 0.1 µg/l, and the bioassay technique simply involved taking subsurface samples from exposed water and observing the effect on test *Gammarus* of 24-h exposures to the samples. Significant mortalities of *Gammarus* were recorded in this way in only two of the five sampling stations, and these only in the period from 1 h to 2 days after treatment.

Studies on parathion in the UK Studies on experimental ponds in which faunal changes following chemical treatment can be interpreted in the light of susceptibility-test data obtained from laboratory tests have also played an important role in a recent evaluation of the toxic hazards of an organophosphorus insecticide of many years standing, namely methyl parathion (Crossland, 1984; Crossland & Bennett, 1984). As in the case of cypermethrin, it was found that immediate and short-term effects of pond treatment were predictable on the basis of laboratory-susceptibility tests, but that longer term or secondary effects occurred which could not have been foreseen. This is well illustrated by the reactions of microcrustacea in treated ponds. Treatment virtually eliminated *Daphnia* populations, with no indication of repopulation until 10 weeks later. There was also a progressive decline in *Cyclops* populations with normal conditions not being restored until about 40 days later. In striking contrast, the methyl parathion treatment produced no reduction in the populations of *Diaptomus*, with no difference between control and treated ponds over the period of 1 week before and 1 week after treatment. From 3 to 13 weeks after treatment, the *Diaptomus* population of the treated ponds actually increased up to the period of 13–16 weeks when the numbers declined to levels below that of the control. One possible explanation of this unforeseen course of events was the decline of the *Cyclops* populations, known to be preferentially predatory on diaptomids, but other contributory factors could have been the absence of competition from the daphnids – highly susceptible to the treatment – or to reduced predation by insect larvae.

On the other hand, this particular study, unlike that on cypermethrin, was not preceded by the same type of detailed laboratory-testing phase involving the local species; the possibility therefore remains that *Diaptomus* was intrinsically less susceptible to methyl parathion, which would account for the sharp difference in their immediate reaction to treatment. In this case the unforeseen development might have proved to be predictable from prior susceptibility-test data. Unfortunately, in the case of parathion, information on this point is too meagre to support or refute this possible interpretation (Mulla & Mian, 1981; Mulla, Mian & Kawecki, 1981).

A somewhat similar situation was observed in the case of Dursban (chlorpyrifos ethyl) applied experimentally to ponds (Hurlbert *et al.*, 1970). Treatment produced a reduction in population of the dominant, but sensitive, species *Cyclops vernalis* and *Moina micrura*, and this was accompanied by the rapid increase from a low level in the population of *Diaptomus pallidus*. Here again, the apparently unpredictable 'secondary effects' could have been the logical consequence of an intrinsically high tolerance to Dursban on the part of *Diaptomus* relative to other species.

Even with the advantage of prior susceptibility-test data, the interpretation of longer-term community changes in aquatic invertebrate populations may still prove difficult; but, in the absence of such basic information, evaluation inevitably moves into the realm of speculation and guesswork.

UPTAKE AND RESIDUES OF PESTICIDES

PESTICIDE RESIDUES IN THE ENVIRONMENT

Ever since the discovery of the remarkably persistent properties of DDT and allied chlorinated hydrocarbon insecticides, and its ability to accumulate in certain living tissues, the problem of pesticides in the environment has been one of the most intensely studied aspects of environmental contamination. This interest which built up so rapidly in the 1960s found expression in the proliferation of new scientific journals to keep pace with the mass of new information. Two of these journals in particular are devoted almost exclusively to residues and to uptake of pesticide by living organisms and components of their environment, namely *Pesticide Monitoring Journal* and *Residue Reviews*; advances are also reported and published in many other international journals as will be seen in the reference list at the back of this book.

In that earlier era of intense interest and progress, one of the most significant findings in the context of aquatic organisms centred, not just on the uptake and retention of DDT by living organisms, but on the biomagnification process whereby uptake and retention of chemical by herbivores and detritus feeders at the base of the food chain was magnified and intensified at each stage of the predator/prey association. In the aquatic medium, the peak of bioaccumulation occurred in fish, but even higher residue levels were attained in fish-feeding birds and their predators in turn. For example, it was found that concentrations of DDT in water as low as 0.05 ppb were able to accumulate through the medium of the food chain to a peak of 75.5 ppm in fish-eating birds (Woodwell, Wurster & Isaacsson, 1967). Invertebrates exposed to concentrations of DDT and aldrin below 0.1 ppb (100 ng/l) were found to accumulate residues rapidly to levels more than 1000 times greater than in the water, thus affecting both the quantity and quality of pesticide residue available to higher aquatic organisms (Johnson, Saunders & Campbell, 1971).

Further support for the food chain concept was provided by studies resulting from pesticide monitoring programmes in the US which indicated widespread occurrence of DDT residues (DDT, DDE and DDD) in both natural waters and in commercial and sports fish. Residues found in Lake Michigan, for example, could only be measured in a few parts per trillion but were found to occur in several parts per million in forage fish. Fish (brook trout) exposed to carbon labelled *p-p'*-DDT in water for 120 days continuous exposure accumulated a total of 3.5% of the total available to them. Under the same conditions, the fish accumulated ten times more from the diet than from the water (Macek & Korn, 1970).

All these events have been so well documented and revived in the many books written on pesticide ecology that there would be no real advantage in covering that familiar ground once more in the present review (Hurlbert, 1975). Such an attempt would also be of doubtful value in view of the fact that since that period of intense interest in DDT, attention has shifted from the persistent chlorinated hydrocarbons such as DDT, dieldrin and aldrin, to the less-persistent allies such as methoxychlor and endosulphan and, to an even greater extent, to the organophosphorus and carbamate insecticides and the synthetic pyrethroids. However, there are some aspects of particular relevance to aquatic invertebrates in general and to stream invertebrates in particular which can usefully be highlighted from that earlier era.

The first significant point is that research showed that the accumulation of such residues in aquatic organisms did not always conform to the clear-cut pattern originally visualised. Among aquatic invertebrates, by far the greatest accumulation of chemical was found to occur in organisms which are not intrinsic components of the food chain or food web, namely the scavengers and detritus feeders such as fresh water mussels. Freshwater mussels, and oysters, placed in pens in river beds in which quantities of DDT and its metabolites occurred in the order of 0.01 ppm were found to accumulate the chemical to very high levels. Exposed continuously for 12 days to 1 ppb DDT, it was found that oysters (even more efficient than mussels) accumulated residues of 14–20 ppm, i.e. a bioconcentration of 14000–20000 times (Bedford, Roelfs & Zabik, 1968).

This was further demonstrated by the model ecosystem technique (see page 113) in which the scavenger mollusc, *Physa* – not in the food chain – concentrated dieldrin (DLN) up to 115000 times, whereas in the fish at the top of the food chain, dieldrin was concentrated up to 6000 times (Sanborn & Ching-Chieh Yu, 1974). In fact this ability of the freshwater mussel to accumulate DDT and DLN has been utilised by state agencies in Michigan, Wisconsin, Minesota and Indiana in the monitoring of pesticide chemicals in tributaries of the Great Lakes (Bedford & Zabik, 1973). Using clam, *Lampsilis*, and mussel, *Anodonta*, in exposures to different concentrations for periods of several weeks, the molluscs were found to accumulate DDT approximately 2400-fold and DLN 1200-fold.

Another observation at variance with the conventional food-chain concept was made in a stream highly contaminated with dieldrin, in which it was found that typical prey organisms such as *Simulium* larvae and hydropsychid larvae were found to take up 12–70 times more chemical than the predator hellgramite *Corydalis cornuta* (Wallace & Brady, 1974). Other investigations too have concluded that aquatic organisms accumulate more dieldrin from the water than from the diet (Chadwick & Brocksen, 1969).

UPTAKE OF METHOXYCHLOR BY MACROINVERTEBRATES

With increasing restriction on the widespread use of DDT, dieldrin and aldrin in insect pest control, more interest was shown in another chlorinated hydrocarbon insecticide, viz., methoxychlor, which had the same desirable quality of low mammalian toxicity shown by DDT but with considerably reduced capacity to persist in the environment. This chemical is of special interest in the context of the present review as it is one which has been very widely used – by direct application to running waters – for control of blackfly larval populations in Canada and to a lesser extent in the US. While most pesticide-monitoring programmes are mainly, if not exclusively, concerned with chemical residues in freshwater fish, the methoxychlor investigations are of particular significance in that uptake by running water invertebrates has been studied at both field and controlled laboratory level.

In the field, residue studies on aquatic insects were made nine days after the application of methoxychlor to the Saskatchewan River in Canada at the rate of 0.309 ppm for 15 min (Fredeen, Balba & Sasa, 1975). No residues were detected in dragonfly naiads or caddis larvae collected from the river bed. However, those insects which had been disabled by the actual passage of the insecticide were found to contain an average of 17.5 ppm methoxychlor; these organisms including stoneflies and mayflies as well as dragonflies and caddis.

In other Canadian studies samples of larvae were caught in nets 275 and 550 m downstream from the point of methoxychlor application. Residues could be detected in *Simulium* larvae at the first of these stations within 15 min of application, and from 30 to 60 min after application at the lower station. Residues ranged from 0.24 to 2.57 mg/l and generally increased with time, up to 180 min at least (Wallace, Hynes & Meritt, 1976)

As part of the same investigation, laboratory experiments were carried out in static tanks with *Simulium* larvae and Trichoptera larvae exposed to insecticide for 15 and 30 min, two formulations being tested, namely an ethanol solution and a particulate formulation. The results Fig. 5.7 show that residue levels for the particulate formulation were higher than those in the ethanol solution with *Simulium*, but the converse was the case with Trichoptera. With *Simulium* larvae concentrations ranged from roughly 20 times (30 min) to 30 times (15 min) greater than the maximum concentration in the water.

In natural running water (Chalk River) to which an oil solution of methoxychlor had been injected, methoxychlor residues in *Simulium* larvae ranged from 240 to 2570 µg/kg, compared to a maximum concentration in the river water of 0.79 µg/l. This suggests that accumulation of methoxychlor may be even more efficient with the oil solution than the particulate formulation used in the laboratory tests, and that this uptake might also be accentuated in running versus static water.

In the *Simulium* control project in the Athabasca River (see page 182) test animals in the form of clam (*Lampsilis radiata*) indigenous to the river, and crayfish, *Orconectes virilis*, which are not, were suspended in the treated river in cages. No mortality attributable to methoxychlor was recorded in the caged animals, but all animals concentrated the chemical above the treatment level, 300 µg/l, with a maximum concentration factor of 33 (Flannagan *et al.*, 1979).

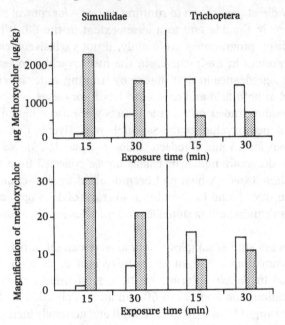

Fig. 5.7. Concentrations of methoxychlor in simuliid and trichopteran larvae after laboratory exposure to particulate formulations (▩) and to ethanol formulations (☐) (Wallace *et al.*, 1976).

More recently the uptake of methoxychlor and another chlorinated hydrocarbon, endrin, has been studied experimentally in the laboratory in a flow-through tank (Anderson & DeFoe, 1980). As the conditions of these tests and the results obtained are of special significance in the context of this review, they are worth examining in some detail. The experiments were carried out in an intermittent flow-through tank. Organisms were exposed for 28 days to five different concentrations of endrin and methoxychlor, the former at concentrations of 0.03–0.6 µg/l, the latter at 0.15–4.23 µg/l, at a flow rate of 22 l/h in each chamber. At the end of the exposure period analysis of residues was carried out by gas chromatograph. Because of the combination of small size and high mortality, no residues were taken in isopods or caddis larvae.

In the case of methoxychlor, residues increased with increasing concentration in stoneflies, reaching a maximum of 1.46 µg/g at a concentration of 4.2 µg/l. If the 'concentration factor' is calculated as the (concentration in animal) divided by the (concentration in water) and based on wet weight, then that figure for the highest concentration in the water is 348. However, the uptake of methoxychlor is proportionately higher at the lowest concentrations, giving a maximum concentration factor of 1130 after exposure to 0.15–0.17 µg/l.

In the case of another test organism, a snail, the methoxychlor concentration ranged from 5000 to 8570 times greater than the water concentration, indicating that the snail bioaccumulates between 13 and 20 times more methoxychlor than the stoneflies at the same water concentration of the chemical.

In the case of endrin, stoneflies exposed for 28 days accumulated the chemical concentrations 600–1000 times greater than the water concentration, following a trend similar to that of methoxychlor.

There are many difficulties in the exact interpretation of these figures in the absence of comparable data. For example, the Athabasca River studies (Flannagan *et al.*, 1979) showed a maximum concentration factor in insects of 33 after exposure to a 15-min pulse of methoxychlor at 300 µg/l, a concentration of 780 times greater than that of the maximum used in the laboratory exposure tests above.

UPTAKE OF ORGANO PHOSPHORUS INSECTICIDES: FENITROTHION

The insecticides which have played a major role in crop protection, forestry and public health in the last 15 years have been mainly organo-phosphorus compounds, carbamates and pyrethrins. Their widespread use is associated with the common property of persisting long enough in the environment to achieve a controlling effect on target fauna, but thereafter to break down quickly with little danger of bioaccumulation in the environment. While the use of these biodegradable compounds is still accompanied by appropriate residue studies in different components of the environment, including (where indicated) measurement of concentration and persistence in water and aquatic organisms, information about aquatic invertebrates themselves is rather scanty. A good example of this is provided by fenitrothion whose ecological impact on forest streams has been the subject of so much study in other directions. Even by the late 1970s, little work had been done specifically on the uptake, metabolism, translocation, activation and detoxification by aquatic invertebrates (National Research Council Canada, 1975). Little was known about the concentration of fenitrothion in aquatic insects following spraying, apart from an isolated record that no fenitrothion had been detected in mayfly nymphs in a brook sprayed at 210 g/ha, a treatment which had reduced the population of these insects (Penney, 1971; Symons, 1977a).

Monitoring records show that in the aquatic environment fenitrothion levels quickly fall off and that even in the more favourable static conditions of ponds fenitrothion has a half life of only 0.3–3.5 days, falling to less than 0.03 ppb by 40 days (Symons, 1977*a*) This latter figure represents the minimum quantity detectable by physicochemical assay (Eidt & Sundaram, 1975) and even in later studies (Morrison & Wells, 1981) it was not possible to detect levels below 0.2 ppb.

However, there is now more direct evidence of organophosphorus insecticide residues in streamsfauna, both in the case of fenitrothion and in acephate (Orthene), one of several insecticides used for control of spruce budworm, particularly in Maine, US (Trial & Gibbs, 1978; Hydorn, Rabeni & Jennings, 1979) where it was the insecticide of choice near major waterways because of its relative safety to aquatic organisms.

In forest streams in New Brunswick, peak fenitrothion measurements in aquatic plants and insects approached 10 ppm after simulated aerial spray, these levels being about 1000 times higher than peak fenitrothion residues measured in stream water (Montreal Engineering Co., 1981).

In the case of acephate, concentrations found in benthic insects in a small coastal stream in British Columbia experimentally injected, were higher than in either fish or sediments (Green *et al.*, 1981). Under experimental 'model ecosystem' conditions (see next section) aquatic invertebrates such as snails can take up significant residues of some organophosphorus compounds, at least under those static conditions (Metcalfe & Sanborn, 1975).

UPTAKE OF CARBAMATES: CARBARYL (SEVIN) AND AMINOCARB

Monitoring studies on aquatic organisms continue therefore to be dominated by residues in fish, and – occasionally – larger freshwater Crustacea. The same general lack of information about the uptake of less-persistent pesticides by aquatic macroinvertebrates applies to the carbamate insecticide carbaryl (Sevin), again the subject of so much ecological study (Mount & Oehme, 1981). Under the favourable static conditions of woodland ponds in forest sprayed areas in Maine, carbaryl residues can persist in pond water and sediments for several months after spraying. Even longer persistence is indicated by the fact that in some organisms reduced numbers may be recorded beyond the year of application (Gibbs *et al.*, 1981). Both those observations probably have little relevance to running water, where again very low concentrations only, down to the minimum detectable concentration of 0.033 ppb, have been recorded (Eco-Analysis Inc., 1982). The minimum detectable level in the pond experiments quoted above was 0.06 µg/l (i.e. 0.06 ppb).

Information about uptake of carbamate insecticides by aquatic macro-

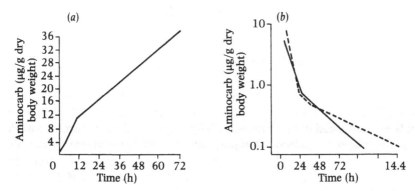

Fig. 5.8. (*a*) Variations in uptake of aminocarb by the aquatic isopod Caecidolea racovitzai with time at an exposure concentration of 1.0 mg/l at 12 °C. (*b*) Clearance of aminocarb by Caecidolea at 12 °C (broken lines) and 20 °C (solid lines) following a 24 h exposure to 0.4 mg/l (Richardson *et al.*, 1983).

invertebrates has been greatly extended by detailed studies on aminocarb and a stream isopod *Caecidolea* (formerly *Asellus* sp) (Richardson, Qadri & Jessiman, 1983). Aminocarb is the active ingredient of the formulation Matacil, widely used for spruce budworm control. The organisms were exposed to 1 mg/l of ^{14}C-labelled aminocarb for periods from 1 to 72 h. In order to examine the effects of concentration and temperature on uptake, isopods were exposed to concentrations of aminocarb ranging from 0.04 to 1.0 mg/l at both 12 °C and 20 °C for a period of 24 h. Additional studies were carried out on the clearance of residues, isopods being exposed to 0.4 mg/l for 24 h, then transferred to clean water and observed for 6 days.

The results of exposure to 1.0 mg/l (Fig. 5.8(*a*)) show that in the exposure period up to 12 h there is rapid uptake as reflected in the steep slope of that part of the curve. Beyond that period, the slope falls off to a steady rate, indicating a continued uniform balance between uptake and clearance. The clearance of aminocarb residues also took a biphasic form (Fig. 5.8(*b*)) in the fast-clearing initial phase, rates were similar at the two temperatures tested, but in the slower clearing phase – subsequent to 24 h post exposure – clearance of aminocarb was proportional to temperature.

In contrast to the detailed laboratory studies, information about residues in the field is relatively scarce. Samples of mayfly nymphs (*Ephemerella*) collected from a book in New Brunswick after an experimental application of aminocarb showed a peak concentration, 1 h post-application, of only 20 ppb, after exposure to a maximum concentration in water of 2.26 ppb, representing a concentration factor of about 9 (Sundaram, Kingsbury &

Holmes, 1984). Thereafter residues declined rapidly to below detection level, coinciding with disappearance of residues in the stream water.

UPTAKE OF LAMPREY LARVICIDE (TFM)

While the bulk of work on uptake of pesticides by aquatic organisms has somewhat naturally been dominated by insecticides, similar problems have not been overlooked in other pesticide fields. This is well exemplified by work on the uptake of the selective lamprey larvicide, TFM, by stream organisms (Maki & Johnson, 1977). This particular investigation is of special interest in that it was based on a model stream system. Each of a series of six indoor troughs, 8 m long and 0.6 m wide, was equally divided into a 25 cm deep upstream pool and a 25 cm deep downstream riffle (Maki & Johnson, 1976a). With continuous flow of water from a natural spring, a natural stream ecosystem was established with natural substrate material and associated flora and fauna. Uniformly C-ring-labelled TFM was used to provide a continuous dilution at the rate of 9 mg/l in the artificial stream, this dilution corresponding to actual field application rates in lamprey control. Continued periodic sampling of the exposed community provided data on retention time and elimination of the TFM by various components.

The TFM concentration in most species showed a rapid accumulation within the first 2 h, after which uptake rates levelled off. This is well illustrated by the figures for the amphipod. *Gammarus pseudolimnaeus* (Fig. 5.9(a)) which follows very much the same pattern as that described earlier for uptake of aminocarb by the isopod *Asellus*.

Studies on the elimination of TFM residues, following a 24-h exposure to the chemical, were carried out on organisms maintained in a flow of clean water. the results obtained with *Gammarus* (Fig. 5.9(b)) showed a concentration of approximately 117 μg/g immediately following the 24-h exposure period, and that half of this original residue – the half-life stage – was eliminated after 26 h. This half-life period was found to range from 7.2 h in the crayfish *Orconectes* to a maximum of 5295 h for annelid worms. A strong correlation appears to exist between the half-life data and the substrate association of the test organisms. Species closely associated with the organic matter and sediments of pool communities, such as the isopod *Asellus* and annelid worms, had significantly longer half-lives than those associated with the gravel and rubble substrates of the riffle community. The mean half-life for pool species was 106 h as compared to a mean of 18 h for riffle-dwelling species.

Among the great deal of material in this investigation, all of which has a considerable bearing on the general problem of uptake of toxic chemicals by stream invertebrates, was the interesting finding that in the caddis larvae of

112

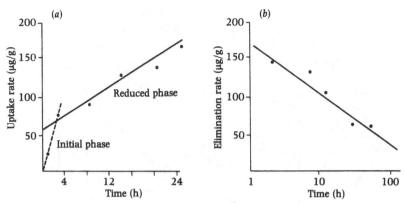

Fig. 5.9. (*a*) Rate of uptake of the selective piscicide ¹⁴C-TFM by the amphipod *Gammarus pseudolimnaeus* during a 24 h exposure to 9.0 mg/l of TFM. (*b*) Rate of elimination of ¹⁴C-TFM by *Gammarus pseudolimnaeus* following a 24 h exposure to 9.0 mg/l TFM (from Maki & Johnson, 1977).

Brachycentrus americanus which has a case composed of vegetable fragments, both the cases and the animals consistently contained high concentrations of TFM.

THE 'MODEL ECOSYSTEM'

The early realisation of the serious environmental effects resulting from widespread use of DDT made it imperative to find alternative insecticides, which would persist long enough to be effective but which would break down eventually and not accumulate in living tissues. It was also imperative to have some advance information about the biodegradability properties of the many new alternative pesticides becoming available. Only in this way could there be any hope of avoiding a repetition of the situation created in the Great Lakes area for example in the 1960s where it was discovered, almost too late to be reversible, that a stable contaminant like DDT could persist for years after it had ceased to be used in major pest control projects in that region; by which time it could accumulate to almost unbelievable levels of 3–10 ppm in trout and salmon from water concentrations as low as 0.000002 ppm (0.002 µg/l).

This urgent need led to the development in the US of the 'model ecosystem' laboratory technique whereby the relevant properties of different pesticide chemicals, including herbicides, could be evaluated by a standardised technique (Metcalf, Sangha & Kapoor, 1971; Metcalf *et al.*, 1973; Metcalf & Sanborn, 1975). The urgent need for such comparable information was also emphasised by the (United States Department of Agriculture) regulation made

113

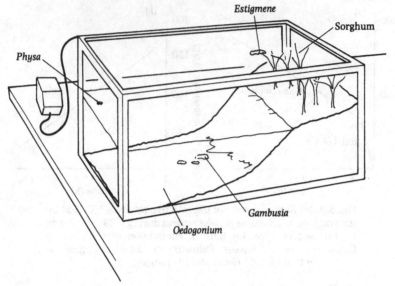

Fig. 5.10. The laboratory 'model ecosystem' (after Metcalfe &
Sanborn, 1975).

in 1969 that such information was an essential requisite for the registration
required for new products.

The basic unit of the model ecosystem (Fig. 5.10) is a glass aquarium
$10 \times 12 \times 20$ inches ($25 \times 30 \times 50$ cm) with white quartz sand moulded into
a sloping surface providing a terrestrial component at one end, and an
aquatic component at the other. Part of the terrestrial end is flattened to
accommodate growing plants, while the aquarium end contains 7 litres of
'standard water' which contains sufficient minerals for the sorghum plants
and for the *Oedogonium* algae in the aquatic section.

The apparatus was also satisfactory for the growth of plankton, *Physa*
snails, Cladocera (*Daphnia*), mosquito larvae and fish. The sorghum plant
provided the ideal host for the second element in the food chain, namely the
salt marsh caterpillar, *Estigmene acrea*.

The actual test procedure was as follows. The ecosystem was allowed to
stabilise for 20 days – after introduction of 10 *Physa* snails and about 30
Daphnia magna – to allow the sorghum plants to grow approximately 10 cm
high. The leaves of the plants are then treated quantitatively – by capillary
pipette – with 5.0 mg radiolabelled pesticide, the rate of application approxi-
mating to field application rates of 1.12 kg/ha (1 lb/acre). Ten early-
fifth-stage larvae of the salt marsh caterpillar are then placed on the treated
leaves. The larvae consume most of the plant surface within 3–4 days, and
their faeces, leaf frass and the larvae themselves contaminate the moist sand
and water. After 26 days, 300 *Culex* mosquito larvae are introduced, and

114

after these have remained in the model ecosystem for 4 days, 50 are removed for analysis of radioactive contaminants. The food chain is completed after 30 days by the addition of three Gambusia mosquito fish which devour the remaining mosquito larvae and *Daphnia*. After three days in the system, these in turn are examined for radioactivity, and at this point – after 33 days – the experiment is terminated by examination of weighted samples of the organisms for radioactivity.

The food chain pathways for the radio labelled pesticide in the model ecosystem are as follows:

1) Sorghum → *Estigmene* (larvae);
2) *Estigmene* (excreta) → *Oedogonium* (algae);
3) *Oedogonium* → *Physa* (snail);
4) *Estigmene* (excreta) → diatoms;
5) diatoms → plankton;
6) plankton → *Culex* larvae;
7) *Culex* larvae → *Gambusia*.

Despite the complexity of this ecosystem, the standardised technique showed that replicates produced very similar results as to distribution and accumulation of total radioactivity in the components of the system, and to the quantitative biodegradability of DDT to DDE, DDD and other metabolites. The tests showed for example that DDT is concentrated and stored in the tissues of the *Gambusia* fish (at the top of the model ecosystem food chain) approximately, 10000–13000-fold over the concentration in the water phase. In contrast, methoxychlor was only accumulated in fish to the order of 0.01% of those found with DDT, and (apart from *Physa*) is not stored in high concentration in living organisms and is environmentally degradable. Associated with these biodegradable properties of methoxychlor is its very much higher solubility in water (500 times more so than DDT).

In addition to other chlorinated hydrocarbon insecticides such as dieldrin, aldrin and endrin, the model ecosystem has provided a great deal of information about other insecticides such as carbamates and organophosphorus compounds which have generally replaced DDT and its allies (Metcalf & Sanborn, 1975), as well as herbicides. For example, the OP compound temephos (Abate) which has been the subject of so much interest in stream and river contamination, appears to possess ideal ecological properties in that it is exceptionally degradable, with no residues detectable in the fish component of the model ecosystem. Even more so was the finding that with malathion – a mosquito larvicide and important contaminant of many static waters – no traces were found in any of the ecosystem components. Parathion, widely used for over 30 years, also showed very little tendency to accumulate in organisms, the fish alone showing detectable residues of the low order of 0.1 ppm.

However, not all the OPs tested showed these desirable qualities; one compound, Leptophos (at one time used for control of cotton pests, but reported to have some undesirable side effects) was found to accumulate in all the organisms of the model ecosystem with ecological magnification (EM) values of 1444 for the fish and 48 398 for the snail. Among the carbamate insecticides, carbaryl is of special interest because of its impact on forest streams in the course of spruce budworm control. This chemical recorded no problems of accumulation in the food chain, and no residues were found in any of the organisms.

As mentioned above, the information provided by model ecosystem data plays an essential part in the registration of new chemicals for general use, and such regulations do have international implications. For use in countries other than the US, the technique is readily adaptable in that different organisms – more suited to the different conditions – can be used as key components of the ecosystem. For example, the model used for some time by the ICI Plant Protection Division uses cabbage plants instead of sorghum, and caterpillars of the tiger moth (*Arctia caja*) instead of the salt marsh caterpillar (Newman, 1976). The mosquito larvae used in this modification are those of *Aedes aegypti*, and the fish component *Lebistes reticulatus* replaces *Gambusia*.

The main value of the model ecosystem has been to provide comparative data, under standard laboratory conditions, on the uptake of different chemical pesticides by different indicator organisms in a terrestrial/aquatic ecosystem. It can also define the paths of chemical transfer, and the distribution and fate of the different metabolites, between the original application point of the chemical to the terrestrial plant and the apex of the food chain as represented by the fish. As the aquatic component of the ecosystem is static, it is difficult to gauge the extent to which these findings can be translated to the very different environment of running water, with its very different range of aquatic faunal components. Obviously, there would be considerable technical difficulties in adapting the model ecosystem concept to one with a running water aquatic component which would appear to allow little opportunity for the chemical or its metabolites to accumulate in the water.

III

EVALUATION IN
PEST-CONTROL
PROJECTS

SIX

INTRODUCTION TO FIELD SAMPLING

OF AQUATIC MACROINVERTEBRATES

IN STREAMS AND RIVERS

INTRODUCTION

In Chapter 1 it was pointed out that many of the striking advances in knowledge over the last 10–15 years regarding pesticide impact have been the outcome of field studies in practical pest control programmes. In all these projects, the environmental studies relate to the known chemical and formulation which is either applied directly to the stream at a predetermined application rate calculated to produce the desired concentration of the chemical in the water, or which contaminates the stream indirectly as a result of aerial application of pesticide to control terrestrial pests in the environs of the stream or river, and where the dosage rate in terms of kilograms per hectare has again been predetermined. In both instances the actual time of application is also known precisely.

The object of these environmental studies is to find out the extent to which the pesticide treatment produces significant changes in the composition of stream fauna produced by mortality or downstream movement, with particular reference to the macroinvertebrates which are the subject of this review. In order to measure these effects and to ascertain significant population changes among the different organisms of running water community – both in the short term and the long term – pesticide ecologists have to rely on a variety of sampling methods. It is these various capture or trapping techniques which provide the essential data for measuring changes in population density or in population composition attributable to pesticide impact.

In the great majority of cases these sampling techniques originate in the well-established methods long practised in pure freshwater biology, and they have consequently been influenced by the considerable advances in that particular discipline in recent years. Noteworthy among those advances is the great mass of new information about natural drift patterns exhibited by

119

many stream invertebrates and the existence of distinct diel periodicity in the benthic drift. The bulk of this knowledge about invertebrate drift has resulted from work in temperate regions such as Europe and North America, but a similar phenomenon has been studied in subtropical areas such as Florida (Cowell & Carew, 1976) and tropical regions such as West Africa (Hynes, 1975). In another direction are the significant discoveries by freshwater biologists that stream invertebrates occur deep in the gravel in the bed of a stream (Coleman & Hynes, 1970; Hynes, 1974; Hynes, Williams & Williams, 1976) and that this 'hyporheic' fraction of the population must be taken into account in assessing the validity of conventional sampling methods.

It is not the intention in this review to discuss the great variety of techniques now available. All that information is already available in fully documented form (Elliott, 1970, 1971c; Hellawell, 1978). For the purposes of this review it will be sufficient to draw attention to the particular sampling techniques which have been most widely used in pesticide-impact studies, bearing in mind the fact that in order to deal with special requirements, pesticide ecologists have on occasions found it necessary to introduce their own modifications of conventional methods, or to devise new and improved sampling techniques. It is for this reason that details of these sampling procedures, and their particular mode of operation, can best be understood and assessed against the background of the particular field operations in which they were used. For introductory purposes therefore it will be sufficient to list the different techniques, leaving the details of their usage to be described in the appropriate projects.

THE SURBER SAMPLER

For sampling macroinvertebrates normally attached to substrates on the stream bed, in rocks, gravel or vegetation, the Surber sampler (Surber, 1937) has been the most widely used technique in studies on pesticide impact. The basis of this method is simply to disturb a standard area of stream bed in such a way that dislodged fauna are collected in a net attached along the downstream edge of the sampling area. The equipment in essence is a combined net and sampling quadrat (Fig. 6.1a), the area sampled being commonly 15 × 15 cm. There is no standard protocol as to the number and disposition of Surber samplers in stream beds, and there is considerable variation from project to project as to number used, number of replicates, disposal across the stream bed, location downstream from pesticide application point, and frequency of sampling. While the technique is suitable for small streams or shallow river edges, the use of this sampler is restricted to

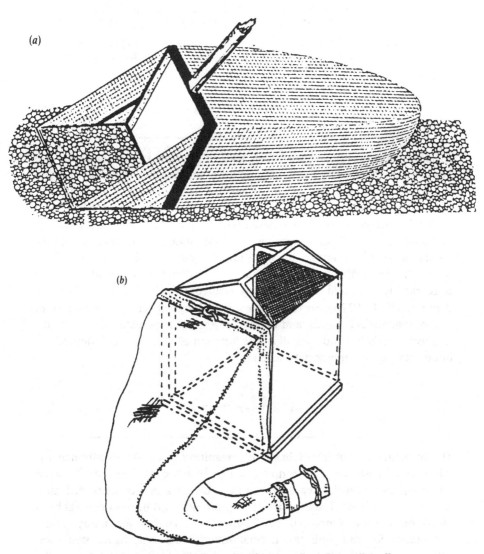

Fig. 6.1. (*a*) Standard design of Surber sampler (after Hellawell, 1978). (*b*) Surber sampler as used on rocky substrates in the Onchocerciasis Control Programme in West Africa (after Elouard, 1984).

running water shallower than arm's length. In some large rivers, such as those involved in the West African Volta River Project, Surber sampling has frequently been carried out on vegetation-covered partly submerged rocky outcrops, often surrounded by turbulent water and accessible only by helicopter (Fig. 6.1*b*).

In stream beds with large stones or boulders making Surber sampling impracticable, the 'kicking' method (Frost, Huni & Kershaw, 1971) has been used on occasions in impact studies as a method of disturbing stream bed organisms and trapping them in a net held downstream from the disturbed section.

ARTIFICIAL SUBSTRATES

Artificial substrates have been widely used in pesticide studies, and their use has been encouraged by the great amount of information available from the extensive studies of freshwater biologists over the years (Williams & Obeng, 1962; Dickson, Cairns & Arnold 1971; Jacobi, 1971; Mason, 1976). These artificial aggregation sites have developed along two rather distinct lines in evaluation studies. Firstly, those specifically designed for the special requirements of *Simulium* larvae in swift-flowing water, and taking the form of cones, plastic ribbons, ceramic plates and other similar substrates likely to be rapidly colonised (Doby, Rault & Beaucournu-Saguez, 1967; Lewis & Bennett, 1974, 1975). Secondly, are those designed for a wide variety of stream macroinvertebrates and taking the form of aggregates of stone balls, 'barbecue baskets', and such like. The function of these will be examined in more detail in the appropriate section.

INVERTEBRATE DRIFT

The measurement of invertebrate drift resulting from, or accentuated by, injection of pesticide into running water has become one of the most important items on the evaluation armament. As a consequence drift nets have been widely used in evaluation programmes, and have in general been based on designs employed by freshwater biologists for many years. Techniques for sampling the normal invertebrate drift have been used extensively over the last 25 years by freshwater biologists ever since the existence of normal drift patterns was disclosed (Waters, 1961, 1962, 1965). There is now a considerable amount of information about the range of equipment used and, even more important, about the interpretation of invertebrate drift as a normal behaviour pattern of the majority of stream macroinvertebrates (Bishop & Hynes, 1969; Waters, 1961–72; Elliott, 1965–71; Townsend & Hildrew, 1976; Allen & Russek, 1985).

DRIFT NET USAGE USING SMALL
TO MEDIUM STREAMS

The methods of sampling invertebrate drift in studies on pesticide evaluation have in general been based on these designs, which usually take the form of a net opening, 15–20 cm square, attached to which is a fine mesh net about 1 m long with apertures in the range 0.4–0.5 mm, adequate to retain the smallest invertebrates but not too small to obstruct the through-flow of water. Two examples of drift net designs used in pesticide studies in Canada, mostly in small streams or creeks, illustrate their practical use. In studies on small shallow streams in New Brunswick, with a stream flow of 17–30 l/s, the mouth of the drift net was 8 cm wide by 40 cm high, with a 66 cm long nylon bolting cloth net with 0.6 mm apertures (Eidt, 1981). The nets were set, usually for 30 min, so that each sampled a column of water from the bottom to the surface.

In similar types of small streams in Quebec, whose width ranged from 2 to 6 m, and depth from 10 to 100 cm, the standard design of drift net measured 0.47×0.32 m, with a 363 μm mesh collecting bag (Kingsbury & Kreutzweiser, 1980*b*). The nets were positioned in such a way that they sampled a water column from the surface to the stream bottom where possible, and from the surface to the net bottom where water levels exceeded the height of the net opening.

In other studies in Canadian streams on the effect of *Simulium* larvicides on drift, rectangular drift nets 929 cm² were used in a fast-flowing stream with a discharge of 2.36 m³/s (2360 l/s) (Wallace & Hynes, 1975), and the identical design was used in other studies on the same problem (Wallace *et al.*, 1973*b*).

In other parallel studies in Canada involving application of *Simulium* larvicide to a somewhat larger rapidly flowing river, the Chalk River, Ontario, with a discharge of 2820 l/s at the time of the trials, pairs of drift nets 30×30 cm and 75 cm long were bolted side by side and attached in mid-channel at each station by clamps fastened to a rope slung across the river, nets being replaced and emptied at 15-min intervals during the experiment (Wallace *et al.*, 1976).

Slightly different designs of drift net were used in *Simulium* larvicide studies in California in small streamlets. In this case the rectangular drift net was 7 cm high and 30 cm wide, with a length of 73 cm (Mohsen & Mulla, 1982). Nets were positioned in mid-stream and drift organisms sampled for 30-min periods. Nets being removed and immediately replaced by new ones at the end of each sampling period.

DRIFT NET USAGE IN LARGER RIVERS

It has long been recognised that the efficiency of conventional drift nets is influenced by many environmental factors such as current velocity, amount of silt or debris in the stream or river water, the exact location of the net in relation to the water column, its siting in relation to the cross section of the stream, and the frequency with which nets are examined and cleared. The most obvious impairment is caused by the net becoming obstructed by silt, or blocked by trapped fauna – of terrestrial as well as aquatic origin. In such cases, strong eddies created at the net mouth offset its function, and also lead to churning and severe damage to trapped organisms.

The effect of such factors tends to be accentuated in the fast-flowing waters of large rivers, and it is in this direction that the basic technique has been subjected to modifications by pesticide ecologists.

In the Onchocerciasis Control Programme (OCP) based on regular applications of larvicide to large West African rivers for control of blackfly larvae (see page 189), sampling of invertebrate drift plays a vital role in evaluation. In the standard method developed, the aperture of the net is 25×25 cm, and the mesh aperture 300 μm (Elouard & Leveque, 1975; Leveque, Odei & Thomas, 1977). In order to provide a large filtering surface, the length of the nets was extended to 2 m. The nets are arranged in several positions across the river – frequently in sets of three (Fig. 6.2) – and located so that the top of the frame is 2 cm below the surface. Samples are taken approximately 1–1.5 h before and after sunset. Measurement of river flow at the time of sampling enables drift to be expressed in terms of animals per cubic metre.

The operation and efficiency of drift nets have been very critically examined by Canadian biologists particularly in connection with pesticide evaluation studies in the large deep fast-flowing waters of the Athabasca River. Prior to the Athabasca studies, which began in 1976, a considerable amount of work had been done in evaluating the effect of *Simulium* larvicides in another large Canadian river, the Saskatchewan. In that investigation attention was concentrated on a variety of artificial substrates designed both for the special requirements of *Simulium* larvae and for more general suitability for a range of non-target macroinvertebrates (Fredeen 1974, 1975). It was not until the Athabasca Project that serious attention was given to overcoming the obstacles in the way of using drift nets in such rapid-flowing rivers in Canada.

It had earlier been recognised that increased net efficiency by reducing clogging of the mesh could be achieved by narrowing the mouth of the net; this is precisely what has been done in designing a new type of 'bomb' drift net (Burton & Flannagan, 1976). The new design is also sufficiently robust to stand up to the exacting physical forces at different depths of the river. A

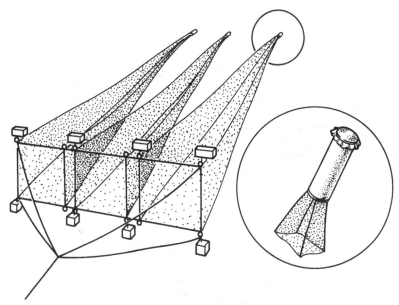

Fig. 6.2. Sets of three drift nets as used in large West African rivers in the OCP (after Elouard, 1984).

simplified diagram (Fig. 6.3a) shows how the mouth of the 1 m long drift 'bomb' is narrowed to a diameter of 15.25 cm by a reversed cone or frustrum of aluminium sheet. The triangular frame which contains the 400 μm mesh net is fitted with three fins for stabilising.

In field trials the bomb design was compared with two conventional drift nets, one a 30 cm diameter round-mouthed, 1 m long 400 μm conical net, and the other a 25 cm square-mouthed 1 m long 200 μm mesh conical net. The results showed clearly that the bomb sampler, in spite of its much smaller mouth opening, was a much more efficient device. The catch of Plecoptera and Ephemeroptera was 3–10 times the net catch, perhaps indicating that the smooth metal frustrum at the front of the 'bomb' net prevents animals from crawling back out after being trapped. The fact that there is also an increase, to a lesser extent, in catch of more sedentary organisms can perhaps be attributed to the improved hydrodynamic properties of the 'bomb'.

In the course of the intensive evaluation studies on the impact of the *Simulium* larvicide, methoxychlor, in the Athabasca River, some modification in the basic design had to be introduced for practical operation in the swift turbulent current (Haufe *et al.*, 1980a). It was found that in the original design with the 15 cm diameter mouth, the intake volume of water exceeded the volume of water that could be discharged through the net. In a swift current, the internal turbulence produced under high fluid pressure, combined with the presence of organic matter and silt, completely pulverises and

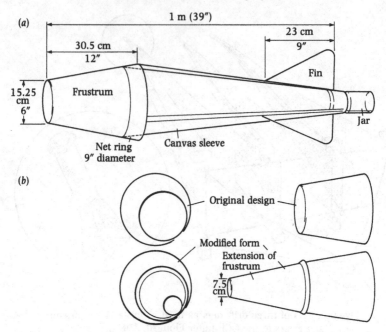

Fig. 6.3. (*a*) A 'bomb' design of drift net suitable for large rivers and fast currents (after Burton & Flannagan, 1976). (*b*) Modified 'bomb' design of drift net, with narrowed frustrum, for fast currents (after Haufe *et al.*, 1980*a*).

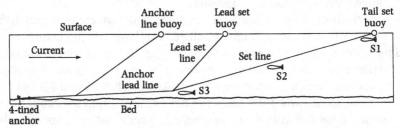

Fig. 6.4. Arrangment of 'bomb' drift nets used in the Athabasca River project, Alberta, Canada (after Haufe *et al.*, 1980*a*). S1, sampler at 50 cm; S2 sampler at mid-depth; S3 sampler at 50 cm above bed.

destroys trapped organisms. The difficulties were overcome by extending the frustrum with the same angle of sheer to a reduced opening of 7.5 cm diameter (Fig. 6.3*b*), reducing the intake volume to 25 % of the original. With a more favourable balance between intake volume and discharge through the net cone, the sampler also maintained a more stable position in even the swiftest current.

Field sampling of aquatic macroinvertebrates

In the operation of these drift nets, sampling at any given point in the course of the river employed two identical sets of equipment. One set was located in the current with the highest velocity in the mainstream. The other was located in the same cross-section of the river, mid-way between this point and the ebb current bank. Each set included sampling devices at three depths; an upper one at 50 cm below the surface of the water: a low level sampler 50 cm above the river bed, and one mid-way between surface and river bed (Fig. 6.4.). All samplers were collected at intervals of 4 h round the clock.

SEVEN

IMPACT OF INSECTICIDES
USED IN CONTROL OF
THE SPRUCE BUDWORM

THE SPRUCE BUDWORM IN CANADA AND THE US

Since the early 1950s the spruce budworm (*Choristoneura fumiferana*) has posed a serious threat in parts of eastern Canada, particularly New Brunswick, (Eidt, 1975, 1977; Symons, 1977a) and adjacent states of the US (Nash, Peterson & Chansler, 1971). In order to protect the valuable timber trees against defoliation, the method of control originally adopted was aerial spraying with DDT, which was practised from 1952 onwards. Since that time the intensity and extent of the infestation has increased. In New Brunswick for example, between 1952 and 1957 the sprayed area increased from 75×10^3 ha to 2.3×10^6 ha (8876 sq. miles). Many of the areas treated twice a year recorded a total application of 560 g/ha DDT per annum, and it was after such heavy treatment that Atlantic salmon, living in streams and rivers in the sprayed forest area, were found to be severely affected.

DDT began to be phased out in 1968, and by 1970 was replaced completely by organophosphorus compounds, mainly fenitrothion. By 1976 the sprayed area in New Brunswick had increased to 4.0×10^6 ha (15000 sq. miles). By that year infestation had extended to other provinces, Quebec, Ontario and parts of Newfoundland and Nova Scotia up to a total area of 30×10^6 ha.

Fenitrothion continues to be the insecticide of choice, in Canada, and at the time of writing appears unlikely to be superseded by other insecticides (D. C. Eidt, personal communication).

128

Insecticides used in control of the spruce budworm

CONTROL BY THE ORGANOPHOSPHOROUS

INSECTICIDE FENITROTHION IN NEW BRUNSWICK

Since the introduction of spraying operations against the spruce budworm in 1952, increasing attention has been given to all the environmental aspects of these operations, especially the effect on freshwater fish and other aquatic fauna of the forest streams (Yule & Tomlin 1971; National Research Council Canada, 1975). As the DDT phase of operations had ended by 1970, and had already been well documented and reviewed (Kerswill, 1967; Muirhead-Thomson, 1971), the present section will concentrate entirely on fenitrothion, and on the continuing work on evaluating its impact on stream invertebrates. The extreme importance attached to the environmental aspects of the spraying operations, by the Canadian government, is demonstrated by the comprehensive range of subjects discussed at a symposium on fenitrothion held in Ottowa in April 1977 (Roberts, Greenhalgh & Marshall, 1977).

CHEMICAL CONCENTRATION PROFILES

There are two important facets to this evaluation; the first, is the chemical measurement of intensity and duration of insecticide impact in these streams according to the different aerial spray regimens in use. Secondly is observation on the effect – both short-term and long-term – of these spraying operations on the populations of stream invertebrates, as judged by drift and bottom-fauna samples. Both of these aspects were examined simultaneously in 1973 at the site of the Nashwaak Experimental Watershed Project in New Brunswick, an interagency project (Eidt, 1975, 1977, 1978). The subsequent course of work on fenitrothion provided a striking example of the extreme difficulties which can be encountered in the way of evaluating the effects of aerial spraying operations on stream biota. Observations had to be geared to the vagaries of the spray programme, schedules which were not modified in any way to accommodate the scientific studies, and consequently there was no way in which a precise experimental approach could be adopted. All of this provides a marked contrast to the much more controllable conditions under which evaluation of another chemical, permethrin, could be carried out in other parts of Eastern Canada a few years later (see p. 140).

The aircraft, flying about 38 m above the tree tops and operating in teams of three, applied fenitrothion at the rate of 210 g/ha. The exact spraying pattern however was subject to a range of variables such as navigational errors, shut-offs over areas of forest declared non-susceptible to spruce budworm, as well as occasional drift of fenitrothion from spraying operations

129

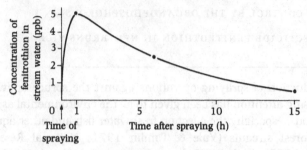

Fig. 7.1. Generalised curve of concentration of fenitrothion in stream water after forest spraying at 140–210 g/ha. (2–3 oz per acre) drawn through median peak concentration, and median times for reduction to 50% and 10% of peak concentration (after Symons, 1977a).

up to a distance of 4 km, and sometimes more, into 'unsprayed' areas. In addition, the extent to which different streams were exposed to the spray cloud was influenced by the extent of tree coverage. Because of all these variables it was not possible in this instance to reach more than a general conclusion.

At a spray rate of 210 g/ha fenitrothion, it can be calculated that if the entire amount settled evenly over a stream whose average depth was 15 cm, the maximum concentration with thorough mixing would be 140 ppb. But, considering the screening effect of most trees, together with the experience gained in previous spraying with DDT, the streams themselves were unlikely to have received more than 25 ppb at most. The chemical assays did not reveal any higher concentrations than 15 ppb.

The three streams under observation showed maxima – within an hour after block spraying – of 5.25 ppb, 1.33 ppb and 6.38 ppb. In the latter stream, the concentration was above 5 ppb for 3 h after completion of spraying. Fenitrothion at concentrations of 2 ppb was still recorded after 8 h in one stream and 12 h in another, thereafter declining to the lowest limit of detection, 0.03 ppb, but could still be detectable for various periods from 24–48 h.

Despite the differences and the variables, these figures provide information of the greatest value in assessing the environmental effect of spraying in that they reveal, not only the actual concentration of chemical to which stream biota are exposed, but also the periods of time over which different concentrations continue to make an impact on stream organisms.

In the present study they had a particular bearing on the parallel investigations, in these same streams, on the effect of spray operations on benthic drift (Eidt, 1975). From all the information available, it is possible to

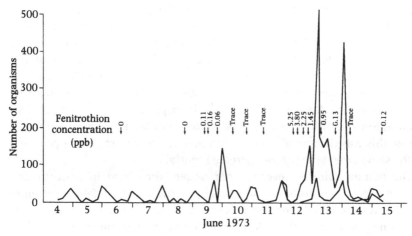

Fig. 7.2. Numbers of organisms in 15 min drift samples from a New Brunswick (Canada) stream – Middle Brook, in relation to fenitrothion concentration in stream water (ppb). The lower line represents living organisms; the upper line represents living plus dead (Eidt, 1975).

construct a general curve of fenitrothion concentration in streams following spraying at 140–210 g/ha (Fig. 7.1; Symons, 1977a).

IMPACT OF SPRAY OPERATIONS
ON FRESHWATER BIOTA

In conjunction with the chemical analysis of fenitrothion in stream waters, observations were carried out on the impact of spraying operations on freshwater biota, particularly in relation to known concentrations of the chemical in the water, and the known duration of impact. The stream invertebrates were sampled both by drift nets and by bottom sampling (Eidt, 1975).

The drift nets, 31 cm wide, were fixed to sample a column of water from the stream bottom, to, and including, the surface. They were set for 15 min every 3 h, each catch being emptied into a holding cage provided with rocks as substrate, and immersed in the same stream for 24 h. The object of this was to check what proportion of the drift, alive at the time of collection, eventually survived or succumbed. The course of events in one of the three study streams is shown in Fig. 7.2.

The drift in normal pre-treatment periods follows a fairly regular pattern, with the greatest catches at night. This pattern was affected both by normal spates and by the presence of fenitrothion. At one extreme an increase in

131

living drift (9 June) was attributed to a fenitrothion concentration of 0.16 ppb 4 hours earlier. At the other extreme, a high drift with a large dead component (12–13 June) followed treatment of the spray block which produced fenitrothion concentrations as high as 5.25 ppb. Dominant among the drift organisms were stonefly nymphs (Plecoptera) with drift increasing 10-fold in *Amphinemura* spp with up to 100-fold in *Leuctra* spp. Of the many other groups, mayflies in particular, the high peaks of drift following peak fenitrothion concentrations, were associated with high mortality. An exception to this was the amphipod *Hyalella azteca* which also showed a peak drift at the same time, but with no increased mortality.

The bottom samples revealed that despite the sharp increase in drift following peak fenitrothion concentrations, the attached fauna remained apparently unaffected. From this observation, and from the fact that the fenitrothion-induced drift conformed to the night maxima of natural drift, it was concluded that the fenitrothion did not penetrate to the bottom substrates. Only when the attached fauna leave their attachment and enter the drift do they come in contact with concentrations of the chemical which – depending on length of exposure – may prove fatal. However, attempts to confirm this concept in the following year (1976) by simultaneously sampling water from the main stream and water from the interstitial spaces in the stream bottom, provided only weak evidence for decreased fenitrothion concentration. This may have been due to technical flaws, or to the fact that the coarse gravel bed in this stream did allow more general circulation than would occur in a stream bed of finer material where one would expect a sharper gradation of chemical (Eidt, 1977).

Up till this point, evaluation of fenitrothion treatment had been greatly handicapped by the vagaries of the aircraft-spraying operations, and by the uncertain degree to which streams in sprayed blocks were exposed to the spray cloud. Considerable progress was made in 1977 when the problem was tackled experimentally by means of injecting known concentrations of fenitrothion directly into the stream itself at a fixed application point. The object was to define more accurately the dose of fenitrothion that would cause drift and 'kill' stream arthropods (Eidt, 1978).

Fifty metres below the application point, a complete bank-to-bank barrier screen was set up to filter out all drift coming from upstream into the test experimental stretch. Tests with Rhodamine B had shown that the insecticide became thoroughly mixed. Drift nets of the design previously used, sampling a column of water from bottom to surface, were set up at various sampling stations from 58 m to 145 m below the application point. Samples were taken at half-hourly intervals, beginning with 30 min before application to $1\frac{1}{2}$ h afterwards. In addition, a drift net sampled the normal drift in the untreated section 12 m above the application point. 4–5 days before treatment began, cages containing representative live stream organisms, were established at

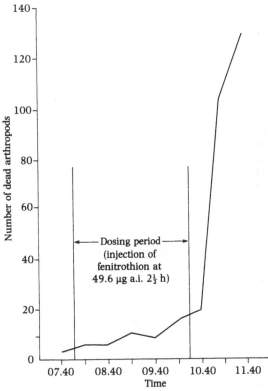

Fig. 7.3. Dead aquatic arthropods collected in 15 min drift samples during and subsequent to a $2\frac{1}{2}$ h dosing period in a New Brunswick stream (totals of 4 stations between 59 m and 146 m below injection site). (after Eidt, 1978).

the 5 downstream sampling stations, on the bottom and at planned depths of 10, 20 and 30 cm in the stony rubble in the stream bed.

A first injection of calculated concentration of 49.6 µg/l active ingredient was followed 14 days later by a dose of 69.1 µg/l, each injection lasting for a period of $2\frac{1}{2}$ h. These two experiments were designed to yield much more precise data than could be provided on the basis of variable aircraft application. However, the smooth running of the experiment and the interpretation of the data were handicapped by two factors. In the first test there was an interruption in the flow of injected chemical of unknown duration in the first 50 min of the $2\frac{1}{2}$ h injection period. In the second test, injection went smoothly, but – unlike the first test – water was not sampled for fenitrothion concentration, nor were drift samples taken. Allowing for the interrupted injection period, the results of the first test were reflected in sharply increased drift rates of dead organisms at all sampling points downstream, starting about 15 min after the end of the $2\frac{1}{2}$ h injection period,

133

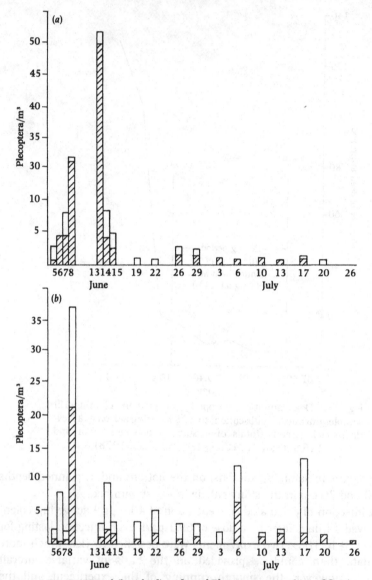

Fig. 7.4. Dead (hatched) and total Plecoptera per m³ in a 30 min net set taken between 21.30 and 22.30 on various nights. (*a*) In stream injected with fenitrothion at 52 µg/l over 1 h 30 min on 6 June, and 73 µg/l over 2 h 10 min on 13 June. (*b*) Untreated control (after Eidt, 1981).

and reaching a peak within an hour (Eidt, 1978; Fig. 7.3). The bulk of dead drift was composed of stoneflies (*Nemoura* sp) followed by mayflies, *Baetis*. These results were produced by a fenitrothion concentration of approximately 20 µg/l for 4 h (the measured concentration being lower than the calculated injection rate – 49.6 µg/l – due to the temporary failure of the injecting equipment). This was not sufficient to kill the main test insects (the stonefly *Leuctra*) in the cages; but the higher dosage in the second test did produce a lethal effect, but this diminished rapidly downstream.

These experiments were repeated in 1978 with the following modification. Rhodamine WT dye was injected along with the dosing equipment (Mariotte bottle), and artificial substrates for sampling benthos were installed in the stream in the form of 'rock balls', i.e. crushed rocks tightly packed in nylon fish net, which were placed in the stream one month before the test started. Two treatments were carried out a week apart at calculated fenitrothion concentrations of 52 µg/l and 73 µg/l respectively. The first injection – over a $1\frac{1}{2}$ h period – took place in the evening (20.00–22.30 h) in order to take advantage of the normal high night-time drift rates. The second injection took place around midday for a duration of 2 h 10 mins.

The first treatment had no immediate effect on invertebrate drift or on standing crop, but on the following three nights there was a larger proportion of dead invertebrates in the drift. The second treatment resulted in a substantial kill of arthropods which began to appear in drift about 15 min after the 2 h 10 min injection period ended. Most of these were Plecoptera and, of these, the majority were *Leuctra* spp. These species also showed a reduction in the standing crop samples, but recovery was complete 50 days after treatment. As in previous experiments, the drift of dead arthropods occurred only on the day of treatment, and not on subsequent days (Eidt 1981, Fig. 7.4).

INVERTEBRATE DEPLETION
AND RETURN TO NORMALITY

In trying to gain a true picture of the short-term and long-term effects of fenitrothion on exposed streams, controlled tests with injections of measured concentrations have played an essential role. However, in some ways, such tests may be unnatural in that the insecticide is injected at one point in the stream, where concentration will always be highest, while downstream it diminishes due to dilution, dispersal and decomposition etc. In contrast, in operational spraying over forest areas, the insecticide cloud may enter a stream over much of its surface and make simultaneous impact over long stretches of the water course. The Canadian workers have been particularly concerned with obtaining precise information about the reduction in insect biomass following spraying and also about the time that elapses before return to normality. They have also given a great deal of attention to trying to estimate at what point the reduction in stream benthos begins to have an adverse effect on growth of fish, as well as the extent to which opportune feeding fish are affected by the short-term high drift rates of poisoned invertebrates following spraying.

Further stream tests with nets for sampling drift organisms and with rock ball samplers of the 'barbeque basket' type recessed into stream beds (Eidt, 1981) have shown that fenitrothion injected into a stream at 73 µg/l diminishes to 23 µg/l at a distance of 343 m downstream, producing a decline in standing crop between this point and 375 m downstream. The same stream was injected on two occasions, the first at a calculated rate of 51.8 µg/l had no immediate effect on either drift or standing crop. This was followed 7 days later by an injection at the rate of 73.0 µg/l which produced a substantial kill of arthropods which began to appear in the drift 15 min after the end of the 2 h 10 min injection period. Again most of the invertebrates affected were Plecoptera, mainly *Leuctra* spp. This reduction in standing crop, as compared to an untreated control, was still evident after 31 days, but had disappeared after further sampling 50 and 357 days after treatment.

EFFECT OF INVERTEBRATE REDUCTION
ON SALMONID FISH

With regard to the question of the effect of the drastic reduction in numbers of stream invertebrates following spraying with fenitrothion on the salmonid fish – Atlantic salmon (*Salmo salar*) and brook trout (*Salvelinus fontinalis*) – controlled laboratory experiments have shown that it is extremely unlikely that aquatic, or terrestrial, insects killed by operational doses of fenitrothion can cause lethal or sublethal effects directly on those fish (Wildish & Lister, 1973). Nor were any fish kills observed in the salmon and trout which had gorged themselves on poisoned insects in a stream treated with fenitrothion at a rate 100 times greater than operational spraying (Symons & Harding, 1974).

It appears therefore that any reported adverse effects on salmonid production can only be due to a sharp fall in the number of fish food organisms following spraying. Because of the wide variations in the rate of invertebrate depletion from stream to stream, it is not possible to establish an exact relationship between spray dosage and reduction in biomass. However, a mass of data from various sources can provide some general indication of a relationship. (Symons, 1977*a*).

Following spraying at 140–210 g/ha, kills of invertebrates vary from no-effect in some streams to 60–70% reduction in some extreme cases; the median reduction of 2–5% is a level of depletion undetectable by present methods of population assessment. At 280 g/ha (4 oz/acre) variation extends from no-effect in some streams to 80–90% reduction in biomass in a number of extreme cases (Varty, 1976), the median reduction being 10%. Only a few streams have been sprayed experimentally at dosages greater than 280 g/ha however the limited data available suggest that at about 450 g/ha, (6 oz/acre) the median reduction in biomass is about 50%, with many streams having 90% reduction or more.

In considering the effect of spraying regimens on fish growth, and relating this to patterns of biomass reduction, the general conclusion is that reduction in growth of salmon is undetectable at the biomass reduction observed at 210 g/ha. At the higher dosage of 280 g/ha, a reduction in the size of salmon at the end of the summer of 13–15% would be expected in those streams which recorded a biomass reduction in excess of 65%. In view of the fact that insect kills in excess of 65% have been recorded on occasions following this high spray dosage, a combination of such high kills and longer-term effects on the invertebrate fauna, could therefore be expected to have a much more severe depression on salmon production than those following the relatively light short-term, effects which were the subject of these estimates (Symons, 1977*a*).

IMPACT OF FENITROTHION IN CONTROL OF PINE
BEAUTY MOTH IN SCOTLAND

One of the most unusual investigations on the aquatic impact of fenitrothion spraying used against forest pests has been carried out under very different zoogeographical conditions from those in North America, namely in Scotland. The unusual feature is that existing pesticide-usage regulations in the UK do not permit large-scale aerial spraying with insecticide except under severe emergency conditions. Such conditions were created in 1977 when it was found that outbreaks of the pine beauty moth (*Panolis flammea*) in northern Scotland the previous year had become so serious and extensive that it became imperative to carry out aerially applied chemical control on a scale previously unknown in Britain (Holden & Bevan, 1979). The choice of fenitrothion was mainly dictated by the extensive experience of this insecticide against forest pests in eastern Canada and the northeastern United States. A total of 5000 ha were covered in the 1978 spraying operations, the blocks being mainly scattered in an area 40 km from north to south, and 45 km from east to west, in the eastern half of Sutherland. In addition, an area of 50 ha in southwest Scotland (Dumfries and Galloway region) was completely defoliated and had to be treated similarly. Spraying was carried out with ultra low volume (ULV) aerial treatment at the rate of 300 g AI/ha.

Intensive investigations on the environmental aspects of this aerial spraying were established at an early date; the two aspects of special concern in this review being the fate of fenitrothion in affected streams and the impact on stream invertebrates (Wells, Morrison & Cowan, 1978).

Initially the main sampling methods for invertebrates were drift nets, 50×20 cm, examined every 6 h for a period of 2 days after spraying. The stony and boulder-strewn nature of most of the stream beds made Surber-type sampling unfeasible, and the 'kick' technique of disturbing large stones and catching dislodged fauna in a net was resorted to (Frost *et al.*, 1971). The use of these two sampling methods disclosed the fact that not every species taken in the drift net was recovered in the kick samples, and the kick samples themselves recorded several species never found in the drift material, such as larvae of the alder fly, *Sialis fuliginosa*.

A sharp increase in the number of invertebrates taken in drift nets was observed within a few hours of aerial application with fenitrothion. The drift decreased rapidly as the concentration of chemical fell – from maxima of 18–48 µg/l, to less than 3% of that within 24 h, eventually falling to levels well below 1 µg/l after 5 days. Spraying operations started in 1978 were continued in 1979, when additional sampling methods were used to study invertebrates, namely the use of a Hess-type sampler for bottom fauna, and the use of caged indicator species – mayfly and stonefly nymphs (Morrison & Wells 1981).

In 1979 a critical study was carried out on an experimental stream whose headwaters were within an area sprayed on 10 June of that year (Morrison & Wells, 1981). The invertebrate-sampling procedures provide an interesting comparison with those devised by the Canadian workers in their studies on the impact of fenitrothion and permethrin on stream fauna (page 131). The experimental stream selected varied from 2–4 m in width over much of its length, and much of the bed consisted of large stones 20 cm or more in diameter. Three sampling stations were selected; the first was just below the treatment area where fenitrothion at the rate of 300 g/ha was applied by fixed-wing aircraft; the second and third stations were 1015 and 2030 m downstream respectively.

Drift net samples were collected over a 4 h period prior to spraying. After spraying, drift nets were sampled, emptied and re-set every 4 h for the first 24 h, then for a 4 h period on the 3rd and 5th day. At three other points between stations 1 and 3, a Hess sampler was used to remove animals from 6 different known areas (0.1 m^2) of the stream bed, immediately before and 9 days after spraying. Throughout the sampling period, caged invertebrates at stations 1 and 2 were inspected at regular intervals.

In the study of invertebrate drift, several of the samples were emptied directly from the net into a shallow tray and observed for a period of 1 h. During this hour, many of the drift organisms were seen to be moving actively, indicating that although they may have been dislodged by the

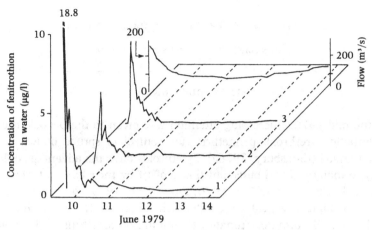

Fig. 7.5. Concentration–time profiles for fenitrothion at 3 stations in a Scottish stream treated on 10 June at 300 g/ha by aerial spraying (Morrison & Wells, 1981).

insecticide, they had not been immediately killed by it. This conclusion was supported by the finding that caged aquatic insects remained alive during the 5-day post-observation period.

As one of the principal objectives of this field study was to monitor the input of fenitrothion into the water courses, and its fate during the post-spray period, valuable information was obtained about the relation between drift and the actual concentration of fenitrothion in the water. The chemical analysis data enabled concentration–time profiles to be charted (Fig. 7.5, Morrison & Wells, 1981) for each of the three stations. The maximum recorded concentration of fenitrothion of 18.8 µg/l was reached at station 1 between 1 and 2 h after spraying. The concentration fell rapidly to below 1 µg/l within 12 h, then decreased more slowly to below 0.2 µg/l during the following 3 days. The movement of the fenitrothion slug downstream is reflected in the time taken to reach maxima at stations 2 and 3, namely 3 and 4 h respectively, and by the concentration–time profiles becoming successively broader. Observations in another river in the previous year (1978) had shown that no obvious drift was produced at fenitrothion concentrations of 1.16 µg/l, while in the present study significant changes took place at concentrations 2–4 µg/l. It is suggested that a concentration zone as low as 2–4 µg/l is sufficient to initiate drift on the part of some invertebrate species.

INSECTICIDES IN SPRUCE BUDWORM CONTROL:

EXPERIMENTS WITH PERMETHRIN

INTRODUCTION

From the mid 1970s, increasing attention was given to the possible use of the synthetic pyrethroid, permethrin, for control of spruce budworm in eastern Canada (Kingsbury, 1976a). Pyrethroids have the advantage of low toxicity to mammals and birds, but this is offset by their known toxicity to fish (Mauck, Olsen & Marking, 1976; Zitko *et al.*, 1979; Zitko & McLeese, 1980). The high insecticidal properties which render them effective pest control chemicals, also unfortunately extend to non-target insects including many aquatic forms (Mulla *et al.*, 1975, 1978, 1980; Muirhead-Thomson, 1978a).

The gradual introduction of permethrin into spruce budworm control was therefore made in the full knowledge that spray operations might produce environmental hazards according to the application rates, and that the fish and insect fauna of water bodies were likely to be the most seriously affected.

Among the first studies to test this were those carried out in lakes and streams in the Forest Experimental Area near Chalk River, Ontario (Kingsbury, 1976a). In conjunction with trials against the prime target, spruce budworm, field trials were also carried out on the direct effect of aerial spraying – at several dosage rates – on the aquatic fauna. The results obtained in treated lakes and streams were compared with those from untreated controls. The experiments on streams, which are the main concern of this book, were carried out in the first instance in a 5 km length of creek which runs into the Ottawa river. The creek varied from 3 to 10 m in width, and from 30 cm to 1.5 m in depth. Applications by fixed-wing plane (Cessna) were made at the rate of 70 g AI/ha directly over the study area, using spray formulations which were dyed with a marker to facilitate measurement of insecticide deposit. Insecticide deposits in the treated stream were estimated by two methods, firstly, by colorimetric measurement of dye deposited on aluminium pans placed at intervals along the length of the creek, and secondly, by using sample cards in association with these pans for counting spray droplets.

In the streams, bottom-dwelling organisms were sampled by Surber samplers, while drifting invertebrates were sampled by two drift methods. In the first, a net 46 cm (18 inches) wide and which stretched from surface to bottom, was set up in the creek for a 30-min period each morning and evening. Immediately prior to, and following the insecticide application, two additional 1-h drift collections were made in the same way in order to study the immediate effects on aquatic organisms, as well as to note the terrestrial

140

insects and other arthropods knocked down by the spray. A blocking seine net set across the entire width of the stream was also installed, and emptied each morning and evening.

Once again it must be pointed out that these stream and creek investigations formed part of a much wider environmental study which embraced studies on fish and invertebrates in lakes, and involved techniques not appropriate to running waters, for example zooplankton collections, the use of emergence traps, studies on fish-stomach contents in relation to insecticide treatment, and the use of scuba and other underwater gear to study fish at aggregation sites.

In the 5 km experimental stretch of the creek, aerial spraying in two identical swathes was carried out along the valley in an upstream direction Over a period of 10 min to produce a total application of 70 g AI/ha (1 oz AI/acre) under stable conditions. Measurements of insecticide deposit showed considerable variation and averaged only about one-fifth of the emitted dosage. As soon as treatment began there was a massive increase of invertebrates on the stream surface, composed of both terrestrial 'knockdown' and of aquatic fauna. Drift net samples which, prior to treatment, were composed almost entirely of normal drifting blackfly and midge larvae, showed a wide variety of aquatic insects immediately following treatment. These numbers remained high during the morning following treatment on the previous evening (19.35–19.45 h), but by the following day had fallen to very low levels.

A similar pattern of drift was also recorded in the seine net collections, which showed in addition a heavy knockdown of terrestrial insects lasting up to 2 days. On the other hand, collections of bottom fauna such as nymphs of mayflies, stoneflies and dragon flies, as well as larvae of caddis flies and water beetles, showed no obvious differences between the treated stream and the untreated control.

Bearing in mind the fact that deposit measurements showed that the actual amount of permethrin which reached the water surface was 13.5 g/ha – only one-fifth of the calculated application rate of 70 g/ha – it was concluded that the experiences in this particular creek could not be taken as a final assessment of permethrin impact. The possibility still remained open that under conditions where a greater proportion of the standard application dosage actually reached the stream, more drastic and longer-term effects might be produced.

EXPERIMENTAL STUDIES IN STREAMS

Following these preliminary experiments in Ontario in 1976, further trials with permethrin were carried out in Quebec in 1978 in three streams forming

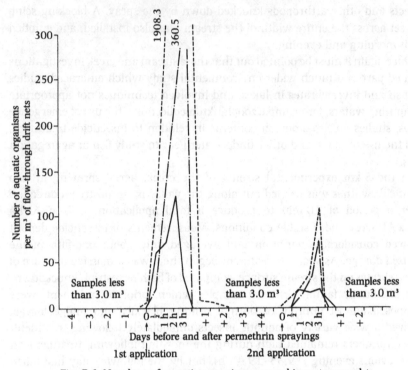

Fig. 7.6. Numbers of aquatic organisms per cubic metre caught in drift nets in 3 Quebec streams (solid line, dashed line, dot-dashed line) treated aerially with double applications of permethrin at 17.5 g/ha at 5 day intervals (after Kingsbury & Kreutzweiser, 1979).

part of the watershed system ultimately flowing into the St John's River (Kingsbury & Kreutzweiser, 1979). Two study streams were treated with a double application of permethrin at 17.5 g AI/ha at 5-day intervals. The width of the treated streams ranged from 4 to 8 m, and the depth from 20 to 70 cm.

Using the same basic evaluation procedures as previously, the results showed first of all that there was again a great variation in the amount of permethrin actually reaching the water surface, from as little as 1% of amount emitted from the aircraft to a maximum of 41%. Analysis of permethrin residues in treated streams ranged from 0.23 to 1.8 µg/l, and fell below the limits of detection (0.2 µg/l) within 24 h.

In all three treatment streams, the initial application of permethrin produced a sharp increase in the number of drifting organisms, and this was still evident up to about 15 h. The increased drift was composed mainly of mayfly nymphs (baetid and heptagenid), stonefly nymphs and caddis larvae (Fig. 7.6). The second permethrin treatment produced a second increase in

Insecticides used in control of the spruce budworm

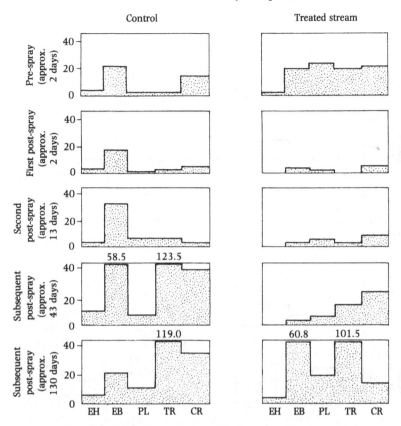

Fig. 7.7. Populations of selected bottom-dwelling invertebrates collected in Surber samples in one of three streams exposed to two aerial treatments with permethrin at 17.5 g/ha; comparison with untreated control stream. EH, Ephemeroptera: Heptageniidae nymphs; EB, Ephemeroptera, Baetidae nymphs; PL, Plecoptera nymphs; TR, Trichoptera larvae; CR, Diptera; Chironomidae larvae (after Kingsbury & Kreutzweiser, 1979).

drift, but smaller than the first. This drift was composed mainly of stonefly nymphs, with caddis larvae noticeably absent.

In contrast to the previous experiments in Ontario, permethrin treatment in this series produced considerable impact on bottom-dwelling invertebrates. This effect was marked after the first application, and intensified after the second, in all three streams. Most affected were those organisms which had composed the drift; heptagenid mayfly nymphs and caddis larvae were virtually eliminated from the stream bottoms, while baetid mayfly nymphs, plecopteran nymphs and midge larvae were reduced by 40–90% in most cases.

The course of subsequent recovery of the treated streams is shown in Fig.

143

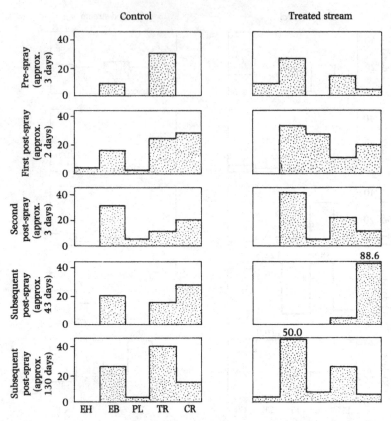

Fig. 7.8. Percentage composition of food organisms in diet of slimy sculpin in one of three streams aerially treated with permethrin at 17.5 g/ha in two treatments; comparison with untreated control. Symbols as for Fig. 7.7 (after Kingsbury & Kreutzweiser, 1979).

7.7 from which it can be seen that most indicator organisms (with the exception of heptagenid nymphs) were returning to normal densities by about 6 weeks after treatment; by the end of 4 months, recovery of all organisms was complete.

Studies on the feeding habits of stream-inhabiting fish provided an unusual insight into another environmental aspect of permethrin treatment, treatment which had no direct harmful effect on the fish themselves. Systematic examination of stomach contents of the slimy sculpin (*Cottus cognatus*) for example, caught by electroshocking, showed that it normally fed on a wide range of aquatic invertebrates. The sharp increase in drift organisms following treatment was reflected in a sharp increased intake of those food organisms by the fish (Fig. 7.8). However, $1\frac{1}{2}$ months later, the diet was found to be composed almost entirely of midge larvae. In the absence of the normal range of food organisms, the reduction in available food was also reflected in one of the treated streams in an increase in the proportion of fish showing empty stomachs. Normally, 17–20% of fish show absence of food in their stomachs, but this proportion increased to between 40% and 60% in the post-spray period.

<h2 style="text-align:center">LARGE-SCALE WATERSHED TRIALS</h2>

Following these experimental studies involving small areas and portions of streams, the next phase of investigation was to study the effect of larger scale applications of permethrin over a much wider area of headwater stream watershed and surrounding forest (Kingsbury & Kreutzweiser, 1980a). The test area selected, in northern Ontario, was 640 ha. Sampling stations were set up in the main stream and two of its tributaries within the treatment block, with an additional sampling station 2 km downstream from the limit of the treated block. Within the treatment block these streams were silty and slow-flowing, interrupted by beaver ponds, and unsuited for Surber bottom sampling. Downstream the velocity increased and there were frequent riffles. Streams ranged in width from 1 m in one of the tributaries to 3–6 m in the main stream. In all of these, physical factors had an important bearing on subsequent evaluation of permethrin impact.

A single application of permethrin was made to the treatment block at the rate of 17.5 g AI/ha. Sampling methods previously used for stream invertebrates were supplemented by two additional techniques, namely, artificial substrates and the use of cages containing known numbers of caddis larvae, dragonfly nymphs and stonefly nymphs, submerged in the stream.

Measurable quantities of permethrin were found in the treated stream immediately after spraying, reaching a level of 2.5 µg/l 1 h post-spray. 2 km downstream from the treated block, permethrin residues appeared 6 h after spraying, reaching a peak of 1.18 µg/l.

Immediately following spraying applications, all sampling stations within the treated block showed an increase in drift, ranging from 2 to 14 times greater than pre-spray figures. The bulk of the drift was composed of stonefly nymphs, insects which had been completely absent from pre-spray samples. Downstream from the treated block, the drift – 12 h after spraying – had increased to a level 110 times higher than pre-spray average, the drift in this case being composed almost entirely of baetid mayfly nymphs and stonefly nymphs. Drift returned to normal after 24 h.

As mentioned above, the nature of the stream beds in the treated block made it impossible to carry out sampling by Surber or rock technique; but downstream from the treated block, where these sampling methods could be used, spraying operations were followed by a noticeable decrease in bottom fauna, which remained low for about 2 weeks. In contrast, none of the caged invertebrates appeared to be affected by the spraying.

In evaluating the impact of permethrin in this semi-operational experiment it appeared that the physical nature of the streams within the treated block were sufficiently different from those used in previous tests to account for certain apparent discrepancies. These streams were much less suited for supporting large populations of aquatic invertebrates, and this was reflected in the low level of pre-spray drift and by the fact that post-spray drift never achieved the high levels previously recorded. These low populations were also reflected in the finding that while certain organisms such as mayfly and stonefly nymphs were prominent in the post-spray drift, they were not present in pre-spray samples.

TRIALS WITH FOUR DOSAGE RATES

During 1977–78, additional field trials were carried out to compare the effects of four different dosage rates of permethrin, namely 70.0, 35.0, 17.5, and 8.8 g AI/ha on streams in two separate areas of the Gaspe peninsula, Quebec (Kingsbury & Kreutzweiser, 1980b). All of these treatments again produced great variations in the amount of each spray formulation deposited at each sampling station, although the emission rates were the same for each application. The amount of deposit reaching the samplers varied from a figure as low as 4.4% to a maximum of 46% of emission volumes.

The insecticide applications at all four dosage rates produced dramatic increases in the number of drifting organisms in the treated streams, averaging from 300 to 700 times greater than the pre-spray drift figures. This was particularly marked in the case of mayflies (Baetidae in three of the four streams, and Heptageniidae in the fourth) followed by caddis larvae, stonefly nymphs and chironomid larvae. The high sensitivity of baetids to all treatments was also replicated in the Surber and rock samples of bottom organisms, mayflies being reduced to the greatest extent in all four treatments.

In the case of the drift records, the magnitude and duration of drift were generally in accord with the application rates. In the case of bottom-dwelling organisms, the differential effect was produced, not so much on the degree of reduction in numbers, but on the duration of effect. This is illustrated for all the invertebrates in Fig. 7.9, and for baetids and mayfly nymphs alone in Fig. 7.10.

In the streams treated at the two lower application rates, sampling stations were also set up in the untreated portion of the streams about 7 km downstream from the treatment area. About 8–10 h after the upper stream had been treated at 17.5 g AI/ha, there was a significant increase in drift in

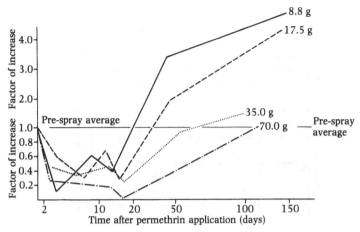

Fig. 7.9. Aquatic invertebrates collected in Surber samples in Quebec streams exposed to aerial application of permethrin at four different rates. Number of organisms expressed as factor of increase or decrease from the mean pre-spray level (after Kingsbury & Kreutzweiser, 1980*b*).

the downstream section, which did not continue however beyond 5 h. At this point downstream the wave of insecticide must have become greatly attenuated, with permethrin concentrations below detectable levels. To offset this, the aquatic fauna would be exposed to these low levels of permethrin for a much longer period – several hours – than in the sprayed area. Consequently, the drift at the downstream stations might not be entirely attributed to drift organisms being carried all that way downstream, but could also be composed in part of local fauna affected at the downstream sampling site. The fact that there was also some reduction in bottom-dwelling organisms at that downstream section would appear to provide some additional support to the idea that permethrin residues had not become completely without effect at that point.

All the streams included in this trial contained abundant populations of brook trout (*Salvelinus fontinalis*) which were unaffected by permethrin treatments. As samples of this species could easily be obtained by electro-shocking and dip net, good opportunities were presented for studying the effect on fish diet of the drastic changes in invertebrate drift and invertebrate fauna following permethrin treatment. Prior to treatment, 75% of the normal diet of brook trout was composed of aquatic insects. Immediately after spraying, the trout consumed large quantities of mayfly and stonefly nymphs, and caddis larvae, provided by the intensified drift. Eleven days after application, terrestrial insects and arthropods made up 80% of the diet, and the greatly reduced aquatic portion was mainly dipterous larvae. This is well illustrated in Fig. 7.11. At the high application rate, 70 g AI/ha, this situation

147

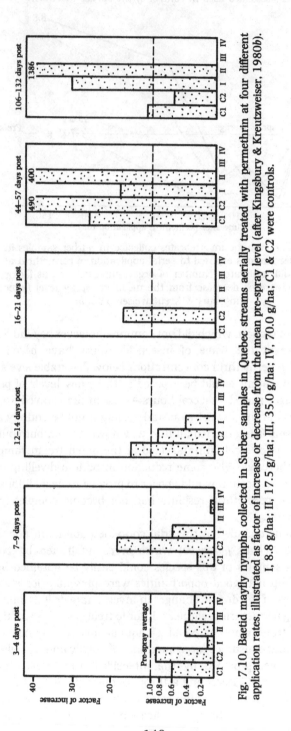

Fig. 7.10. Baetid mayfly nymphs collected in Surber samples in Quebec streams aerially treated with permethrin at four different application rates, illustrated as factor of increase or decrease from the mean pre-spray level (after Kingsbury & Kreutzweiser, 1980b). I, 8.8 g/ha; II, 17.5 g/ha; III, 35.0 g/ha; IV, 70.0 g/ha; C1 & C2 were controls.

Insecticides used in control of the spruce budworm

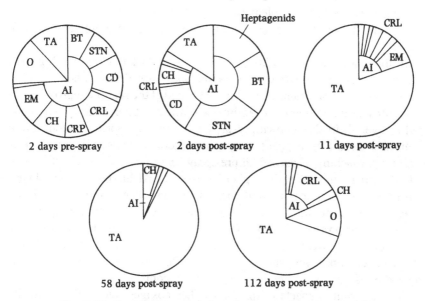

Fig. 7.11. Contribution of various food organisms to stomach contents of brook trout in Quebec streams aerially treated with permethrin at rate of 70 g/ha (after Kingsbury & Kreutzweiser, 1980b). TA, terrestrial arthropods; BT, baetids; STN, stoneflies; CD, caddis; CRL; crane fly larvae; CRP, cranefly pupae; CH, chironomids; EM, emphids; AI, aquatic insects; O, other aquatic invertebrates.

continued for over 3 months. At the lower application rate of 35.0 g AI/ha, feeding on aquatic invertebrates had returned to almost normal levels by that time, but still showed a noticeable absence of mayfly and stonefly nymphs, which, prior to treatment, had provided up to 60% of the total diet. At the application rate of 17.5 g AI/ha, recovery was quicker, and within 8 days the diet consisted of over 75% aquatic insects.

FURTHER TRIALS AT TWO HIGH DOSAGE RATES

In order to obtain further information about the extent and the duration of the two aerially applied experimental permethrin treatments at the higher dosages of 35.0 and 70.0 g AI/ha, the studies which were initiated in 1977 were continued in the same streams through 1978 to 1981. These were the three cold-water streams in the Gaspe peninsula, Quebec, flowing north into the St Lawrence River, two of these being used for the experimental treatments and one as untreated control (Kreutzweiser & Kingsbury, 1982). In addition to the sampling methods already described, the assessment of bottom fauna was supplemented by collections of invertebrates from four randomly selected rocks – approximately 15 cm in diameter – at each station.

At the lower dosage (35.0 g AI/ha) severe depletion of benthos had occurred 3 days after treatment, this being most marked with Ephemeroptera, and to a lesser extent Trichoptera; but by the end of the season, approximately $3\frac{1}{2}$ months after treatment, all benthos (with the exception of Ephemeroptera) had recovered. In the following year, i.e. 1 year post-spray, bottom fauna samples showed a resurgence of Ephemeroptera nymphs and total invertebrates approaching or exceeding pre-spray levels.

In the stream treated at the higher level of 70 g AI/ha the severe depletion of benthos 3 days after treatment was still evident after $3\frac{1}{2}$ months, with the exception of chironomid larvae. These greatly reduced numbers were still evident 1 year post-spray, and it was not until the end of that second year (16 months after treatment) that there was a resurgence of bottom fauna and diversity. Subsequent samples in 1979, 1980 and 1981 showed a return to conditions similar to these in the untreated control.

These differential effects of the two higher dosages were reflected in the diet of the brook trout as estimated from stomach samples. Following the lower application rate, 35.0 g AI/ha, the immediate reduction in the proportion of aquatic invertebrates in the diet recorded, had largely disappeared $3\frac{1}{2}$ months later, although Ephemeroptera and Plecoptera were still noticeably absent.

In the case of the higher dosage, 75 g AI/ha, the diet of brook trout continued to consist almost entirely of terrestrial arthropods for the remainder of the sampling season. In the case of slimy sculpins which occurred only in this higher treatment stream, these were found to rely mainly on Diptera larvae and non-insect aquatic organisms as alternate food sources up to the end of the season. The dependence of brook trout on alternative food sources was still evident 1 year after the application, with a noticeable absence of Ephemeroptera nymphs, and only minor representation of Trichoptera and Plecoptera. During that same period, these three taxa made up 62% of volume of food organisms consumed by brook trout in the untreated control stream. By the end of the following year (1979) and in 1980 and 1981 early season collections, the feeding pattern was similar to that in the untreated control. Despite these drastic changes in diet, and in the availability of normal aquatic food organisms, there was no evidence of pesticide-induced fish mortality or that fish suffered any other harmful effects.

There is some evidence that the slow recovery of benthos in these treated streams may have been accentuated by the virtual elimination of upstream refuges from which drifting invertebrates might otherwise have moved downstream into the treated section. In this trial the flight of the aircraft, during spraying of the main stream portion, continued in an upstream direction as long as the stream was observable from the air, and this could well have eliminated most foci from which rapid recolonisation might have taken place.

IMPACT OF NEW PRACTICAL CONTROL REGIMEN
WITH DOUBLE APPLICATIONS OF PERMETHRIN

From 1979 onwards the course of research on permethrin impact on stream fauna entered a new phase which was determined by the new approach to the practical control of spruce budworm. By that time it had been established that adequate and effective control could best be achieved by employing one of the lower dosages, viz., 17.5 g AI/ha, and by carrying out two aerial sprayings separated by a few days. This new practical phase marked the end of the experimental period on trials at higher dosages.

A series of large-scale trials at this practical dosage were carried out concurrently in three separate provinces of Canada, Ontario, Quebec and New Brunswick. The trials in New Brunswick (Kingsbury, 1982, 1983) are of particular interest in that the spray blocks were intentionally selected to contain portions of the nursery streams of Atlantic salmon (*Salmo salar*), enabling environmental data on this important fish species to be compared to that previously obtained for brook trout. Trials were carried out in a stream system flowing into the Nashwaak River, an area of New Brunswick where many of the early classical studies on DDT and fenitrothion had been carried out. Two 600 ha treatment blocks were selected. One of these received a single application of permethrin at 17.5 g/ha, while the second was given 2 treatments at this dosage 4 days apart. Unsprayed upstream portions of streams exposed to this treatment were used as control sites.

Intensive studies were carried out on permethrin concentrations in those streams exposed to the single and double treatments. In this connection it is significant that over a comparatively short period, improvements in gas-chromatography techniques had reduced the lowest limits of detection of permethrin from 0.2 μg/l in 1979 to 0.05 μg/l in 1980 and to 0.01 μg/l in the present study (Wood, 1983). This meant that in comparing these results with those of earlier trials, water samples previously reported as containing no detectable permethrin may have fallen within the range of detection now possible. Probably for this reason, detectable permethrin residues in the present trials were still found to exist 48 h after spraying, as compared with a maximum of 24 h in the earlier trials (Kingsbury & Kreutzweiser, 1979, 1980a). The question of relating permethrin concentrations recorded in these streams, to the tolerance levels of different fauna at risk, will be examined in more detail in Chapter 5.

Drift studies endorsed previous experience in showing that permethrin application to the blocks caused massive disturbance of aquatic invertebrates resulting in catastrophic drift for 3–12 h (Eidt & Weaver, 1982, 1984). Despite higher spray deposits in the double-application block, and high permethrin residues after the second application, invertebrate drift did not reach the density found after the first spray, suggesting that the bulk of the invertebrates had been reduced or removed by the first spray.

Analysis of drift samples showed a sequence in appearance of different invertebrates at risk, according to their susceptibility and the exposure of their habitats. Among the first to appear after each spraying were blackfly larvae (Simuliidae) which live in exposed sites in rapidly flowing water, followed quickly by stonefly nymphs (mainly *Leuctra* spp) and mayfly nymphs (mainly *Baetis* spp). All but a few of the invertebrates in post-spray drift were dead.

For the first time in this series of trials, introduction of a new technique in the form of the diving mask enabled the fate of drift invertebrates to be followed by means of direct observation. Visual examination of the stream bed downstream from the double-application block, about 54 h after treatment, revealed large numbers of aquatic insects lying on the stream bed behind rocks and in deep slow-flowing areas. These were mostly stonefly nymphs and caddis larvae, and many of the latter case-bearing species were seen to have crawled out of their cases (see page 87). Most of the insects appeared to be dead or moribund, but among the large number of stonefly nymphs (*Phasganophora capitata*) there were indications of apparent recovery.

Concurrent with the drift samples, regular collections were also made of the bottom fauna near the same site (Eidt, Weaver & Kreutzweiser, 1983). In addition to the Surber sampling in riffle areas, and the rock balls or artificial substrate samplers – as previously used – which were inserted in the stream bed 4 weeks before treatment, an additional method was used specifically for simulid larvae. This took the form of artificial substrates 10 cm × 10 cm of unglazed tile which were found to be rapidly colonised by these larvae (Lewis & Bennett, 1974). Because of their exposed habitat in swift-flowing parts of the stream, as well as their filter-feeding habits, it was expected that the standing crop of *Simulium* larvae would be specially vulnerable to permethrin treatment; this indeed proved to be the case, particularly in the double-application block where the population was reduced by nearly 95% after the second spraying.

Apart from its effect on *Simulium* larvae, the single application of permethrin in this series did not produce a clear-cut effect on other members of the benthos. Only after the double application did stoneflies (Plecoptera) and mayflies (Ephemeroptera) show substantial reductions in both Surber and rock ball samples, with baetids and *Simulium* larvae virtually eliminated.

One important disclosure from these trials was that despite the severe disruption produced on stream invertebrates by two applications of permethrin at 17.5 g/ha following so closely on each other, the overall impact was not so severe or so prolonged as the impact previously described after single applications of permethrin at higher application rates of 35.0 and 70.0 g/ha to trout streams (Kingsbury & Kreutzweiser, 1980*b*, Kreutzweiser, 1982).

With regard to the downstream effect, the sampling station 1.4 km below the treated block showed a similar decline in benthos to that within the treated area. However, as in previous studies at the same application rate in Quebec (Kingsbury & Kreutzweiser, 1980*b*) this effect could not be entirely attributed to downstream passage of permethrin from the treated block, but might also have been brought about by significant contamination through aerial drift. Further insight into this possibility is provided by the trials carried out in Quebec concurrent with those in New Brunswick (Kreutzweiser, 1982).

In the Quebec trials extensive use was made of sample cards to study the spray pattern. These showed that permethrin was still deposited 3 km downstream from the treatment block, confirming significant aerial drift and contamination of the stream well outside the treatment zone. For example, in one of the streams in this series there was a sharp increase in invertebrate drift 3 km downstream from the treated block within 1 h of the first application. The portion of stream between the treated block and the downstream station was swift flowing, with an estimated velocity of 0.5 m/s. At this rate, residual permethrin or passive drift from the treatment block would consequently require a minimum of 2 h to be transported the 3 km. The almost immediate post-spray drift at this downstream station must therefore be attributed to aerial drift and contamination of the downstream portions following aerial spraying operations.

Despite catastrophic drift and severely depleted benthos in all but one of the treated streams in the Quebec trials, repopulation of bottom fauna was evident within $2\frac{1}{2}$ months, and was virtually complete by $3\frac{1}{2}$ months. By 1 year post-spray, conditions had evidently returned completely to normal.

Another illuminating feature of these trials in Quebec was the introduction of live test invertebrates to monitor pesticide impact. The aquatic insects in question were stonefly (Plecoptera) nymphs, which were exposed in small tubular holding cages 30 cm long and 10 cm wide, and screened at both ends. A series of polyethylene baffles were placed inside the cages to provide current eddies for the invertebrates. Nymphs exposed in these submerged cages in treated streams did not die, but they exhibited severe stress and disorientation for a period of up to 48 h, followed by apparently complete recuperation. Observations on other drifting invertebrates in treated streams showed that many of them (Ephemeroptera in particular) displayed bouts of great activity alternating with periods of disorientation and erratic movements. This behavioural reaction to permethrin intoxication has been suggested as a reason why aquatic invertebrates under those conditions of exposure drift for much greater distances than normally occur.

IMPACT OF OTHER INSECTICIDES USED IN SPRUCE

BUDWORM CONTROL

COMPARISON OF CONTROL PRACTICES IN
NEW BRUNSWICK, CANADA AND THE ADJOINING
US TERRITORY, MAINE

The two adjoining territories of New Brunswick (Canada) and Maine (US) have a common problem in controlling the spruce budworm. With the phasing out of DDT which is no longer acceptable, almost continuous trials with new candidate insecticides have been carried out in both countries since the early 1970s. As has been described earlier in this chapter, fenitrothion (Sumithion, Accothion) early became the insecticide of choice in New Brunswick, and continues to be the main insecticide of use in that province (D. C. Eidt, personal communication). In the adjoining state of Maine, DDT was also initially replaced by fenitrothion under the trade name of Accothion in the early 1970s (Nash *et al.*, 1971) but from 1975 onwards the carbamate insecticide, carbaryl or Sevin has been used extensively (Stratton, 1982, 1983, 1984). This has led to fresh field studies on the impact of this chemical on the fauna of forest streams (Lotel, 1977; Hulbert, 1978; Courtemanch & Gibbs, 1980).

In Canada, owing to a different regulatory system, there was a delay in acceptance of Sevin for use on a large scale, and it was not until 1980 that a full-scale experimental trial could be implemented in New Brunswick (Holmes, Millikin & Kingsbury, 1981). Although the field investigations in these two adjoining territories had the same common objective, viz., the impact of the insecticide on the fauna and the ecology of forest streams in sprayed areas, the work was carried out quite independently with a significantly different approach to sampling techniques and their evaluation. A comparison of these two main trials, using essentially the same oil-based carbaryl formulation (Sevin-4-oil in Maine and Sevin-2-oil in New Brunswick) provides lessons of considerable significance in the wider field of pesticides and stream invertebrates.

TRIALS WITH CARBARYL(SEVIN) IN MAINE

The object of the Maine trial carried out in 1976 (Courtemanch & Gibbs, 1980) was to examine both short-term (within 60 days) and long-term effects which persisted into the following year. Three streams in each of three different treatment areas were monitored as follows: streams in areas sprayed with

Insecticides used in control of the spruce budworm

carbaryl(Sevin) for the first time, at the rate of 840 g AI/ha, and called '1-year studies'; streams in an area treated at the same rate as above in the year of study, which had also been treated the previous year at the rate of 1120 g AI/ha, and referred to as '2-year streams'; and three control streams which had never been exposed to chemical.

Invertebrate drift samples, collected by means of surface to bottom nets, were taken at 14 and 2 days before treatment, and at 2, 30 and 60 days after treatment, samples being collected for 24-h on each occasion. Drift was expressed as the 'number of organisms collected in 24-h'. Drift samples were sorted in the field to separate live from dead specimens. Surber samples of bottom fauna were taken 14 days before treatment, and 1–3, 30 and 60 days after treatment.

TRIALS WITH SEVIN IN NEW BRUNSWICK

In the New Brunswick trials in 1980, carbaryl (in the form of Sevin-2-oil) was applied as a 'split-application', the insecticide being applied twice to a 400 ha spray block, with a 6 day interval between treatments. The rate of treatment – 280 g AI/ha – was much lower in the Maine trials. Aquatic invertebrate drift was sampled by means of surface-to-bottom nets, with collections of 15 min duration each morning and evening for 9 days before the first treatment and up to 5 days after the second application. Records of current velocities at drift net sites enabled invertebrate drift to be expressed as 'the number of organisms per cubic metre of water in the drift column'. All drift organisms were preserved, with no attempt to sort out live from dead. Bottom fauna were sampled both by Surber sampler and by artificial substrates in the form of rock balls of crushed rock.

The impact of the Sevin treatment on invertebrate drift in these two trials is shown in Figs 7.12 & 7.13. These show that in the Maine trial there was a 170-fold increase in drift 2 days after treatment of the 1-year streams, and virtually all organisms in the samples were dead. Approximately three-quarters of the peak drift comprised stonefly nymphs (mainly *Leuctra* spp), the balance being mainly composed of mayfly nymphs (Ephemeroptera). In the 2-year streams, increased drift after treatment occurred in two of the three streams, 95% of the drift organisms being *Simulium* and other Diptera larvae. Here again, all organisms were dead. In the 30- and 60-day samples, drift had fallen to a low level.

The benthic samples in the Maine trials showed that within hours after treatment of the 1-year streams, many large mayfly and stonefly nymphs were found dead in the stream, with case-bearing caddis larvae distressed to the point of leaving their cases.

In the 1-year streams, the drastic effect on these invertebrates was reflected

155

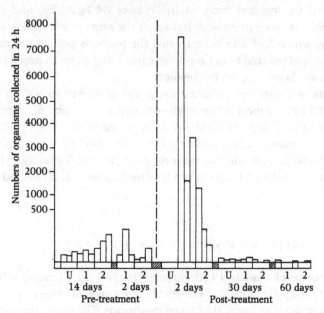

Fig. 7.12. Effect of aerial application of carbaryl (Sevin-4-oil) on invertebrate drift in treated streams in Maine, USA, according to different spray histories. Comparison with untreated control (after Courtemanch & Gibbs, 1980). U, untreated, 1, treated 1 year; 2, treated 2 years. Note discontinuous scale.

in the 30- and 60-day samples which showed that stoneflies (Plecoptera) had been reduced to zero, and mayflies (Ephemeroptera) still significantly reduced. The likelihood that the effect on stonefly nymphs persisted beyond that period is indicated by the fact that in the 2-year streams, only low initial populations of stonefly nymphs were found.

In the New Brunswick trials, a rather different drift pattern was disclosed following the split treatment with Sevin at 280 g AI/ha. Two peaks in aquatic invertebrate drift were observed (Fig. 7.13) following the first application; the first one half an hour after application, and the second one $3\frac{1}{2}$ h after application. The invertebrate composition of these two peaks was quite different, with *Simulium* larvae making up 83% of the first peak, showing a drift rate 145 times greater than pre-spray. In the second peak, Ephemeroptera (baetids) made up 89% of the total, with a drift rate 648 times pre-spray figures. Stonefly nymphs, scarce in the first peak, made up 7% of the second peak.

These three drift components, *Simulium*, *Baetis* and Plecoptera were still drifting in abnormally high numbers 6 h after application. By the evening, drift of all invertebrates had returned to normal. Following the second

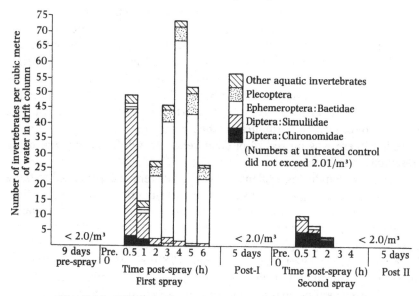

Fig. 7.13. Effect of split application of carbaryl (Sevin-2-oil), applied twice at the rate of 280 g/ha at a 6 day interval, on aquatic insects collected in drift nets in treated streams in New Brunswick, Canada. Number of drift organisms expressed per cubic metre (after Holmes *et al.*, 1981).

application of Sevin, the drift pattern was much less pronounced and showed only a single peak.

At the peak of impact of insecticide following the first spraying in the New Brunswick trials it was estimated that approximately 75 000 aquatic invertebrates drifted past the sampling station. In the 6 h immediately following the first spray however (in contrast to the experience in Maine) there was little evidence of any consistent reduction in standing crop, and no obvious reduction in numbers taken in artificial substrates and rock samples.

ASSESSMENT OF DIFFERENT EXPERIENCES IN MAINE AND IN NEW BRUNSWICK

Perhaps the simplest explanation of these different experiences lies in the wide difference in application rates, the higher application used in Maine being well above the lethal threshold for most aquatic invertebrates, while the lower New Brunswick dosage rates were closer to the no-effect level. It is interesting to speculate whether the two distinct drift peaks recorded after the lower New Brunswick treatment were a manifestation of this borderline effect, or whether two drift peaks, perhaps at shorter intervals, might also

157

have occurred after the high Maine treatment but been masked by the fact that the first drift sample was made 2 days after spraying, and not (as in New Brunswick) by two daily collections from the first post-spray day onwards.

If the drift pattern in the New Brunswick trials is associated with a borderline concentration of carbaryl, then it would be important to know the extent to which drift might still have been induced by sublethal levels of the chemical; in which case a significant proportion of drift organisms would be expected to recover. Unfortunately in the New Brunswick trials, unlike those in Maine, there was no record of dead versus live drift, all material being preserved on collection.

A final and convincing explanation of the differences experienced in these two sets of trials might be expected from the abundant records, in both territories, of carbaryl concentrations in the experimental stream waters following spraying operations. In Maine for example the results from several different trials using applications at the higher rate of 840 g AI/h (Hulbert, 1978; Gibbs *et al.*, 1979; Stanley & Trial, 1980) showed concentrations of carbaryl ranging roughly from 20 to 40 ppb, and even lower levels, below 8 ppb, in streams protected by a buffer zone. In contrast, the monitoring in the New Brunswick trials showed very much higher levels than those above, namely 314 ppb after the first spraying and 123 ppb after the second spraying, even though application rates were 3–4 times lower. It seems possible that this apparent discrepancy could be attributed to the fact that in the New Brunswick trials, carbaryl measurements were made within a few minutes of direct aerial application of the insecticide (Holmes *et al.*, 1981). Within a half to 1 h after spraying, carbaryl concentrations had fallen to roughly the same range as recorded in the trials in Maine (and in Washington State). Another contributing factor may have been the different mode of collecting the water sample, according to which portion of the water column was sampled. Immediately after spraying, an insecticide formulation in oil would tend to concentrate near the surface, causing samples from the top 1 cm of the water column to give unrepresentative values for the stream as a whole, and vice versa.

IMPACT OF ADDITIONAL CHEMICALS USED
IN SPRUCE BUDWORM CONTROL

In addition to fenitrothion and Sevin, several other chemical insecticides have been used over the years in spruce budworm control on a sufficiently large scale to prompt environmental studies, including those on aquatic life. Experimental trials with new promising insecticides or new formulations also demand field studies on their impact on stream ecology. Consequently, there is now a great deal of information about the effect of the following, in particular: the carbamate insecticides Zectran (mexacarbate) and Matacil (aminocarb), and the organophosphorous compounds Orthene (acephate), Reldan (chlorpyrifos methyl) and azamethiphos.

Rather than attempt to cover these insecticides in the same detail given to fenitrothion, permethrin and Sevin, perhaps a more useful purpose would be served by selecting special features of these additional investigations which are of particular significance in the general context of this review.

In the case of the Matacil trials in New Brunswick, for example (Holmes & Kingsbury, 1980, 1982), a new problem was encountered by the disclosure that a Matacil formulation used for some time in the field was more toxic to fish than the pure ingredient, aminocarb. Further laboratory studies confirmed that the so called 'inert' formulating chemicals possessed biocidal properties, the chemical in this case being identified as nonyl-phenol (McLeese *et al.*, 1980*a*,*b*). In order to separate the environmental impact of these two compounds, field trials had to be designed to compare the effects of the Matacil formulation with that of the nonyl-phenol alone, applied at precisely the same rate as in the complete formulation (Holmes & Kingsbury, 1980). These careful trials however, involving drift netting and bottom faunal studies, did not resolve the problem completely, as the nonyl-phenol on its own did not appear to have any significant environmental effect on aquatic invertebrates. As far as salmonid fishes were concerned there was a wide margin of safety between the actual concentrations of nonyl-phenol in flowing water and the lethal thresholds to these fish.

In further field trials involving hand application of chemical to test streams, new formulations of Matacil containing formulating chemicals other than nonyl-phenol were tested against the main 'inert ingredient' (Holmes & Kingsbury, 1982) but conditions were not suitable for conclusive results to be obtained.

There still remains the possibility that although field trials may fail to reveal any clear biocidal action on the part of the suspect 'inert' ingredient, it may still (when formulated) produce a sufficiently potentiating effect on the active ingredient as to account for the enhanced lethality of the whole formulation. There is also an alternative possiblity that an otherwise inert ingredient might produce a formulation which has a greater tendency to remain on the surface of the stream in the form of an 'oil-slick', thus increasing the exposure of surface-dwelling organisms, or physically trapping them.

The really significant point which emerges from these comparison trials is that what appears to be a decisive field trial to compare inert filler alone with the complete formulation containing the active ingredient, may still fail to produce a clear-cut result due to a variety of physical and chemical factors affecting the experiment.

In the case of treatment with the organophosphorus compound Reldan (Kingsbury & Holmes, 1980; Holmes & Millikin, 1981) the particular feature selected for discussion is the attempt by those authors to examine their field data in the light of available information about the precise tolerance levels of different stream invertebrates based on laboratory-simulated-stream test procedures (Muirhead-Thomson, 1978c). In the absence of such laboratory data, based on running water tests, for Canadian aquatic invertebrates, the only comparison possible was with closely related European counterparts. Despite these inherent differences, the analysis proved to be illuminating, but will be discussed in more detail in a more appropriate section of this book.

In the case of trials with azamethiphos (Kingsbury, Holmes & Millikin, 1980), the special point of interest selected was the finding that after two consecutive aerial sprayings at 0.07 kg AI/ha, the general effect on aquatic invertebrates was minimal, with the notable exception of caddis fly larvae of the family Philopotomidae, which showed a large increase in drift, forming in fact 94–8% of the total drift. The apparently selective action of the chemical on this family alone must be related to the fact that these larvae are net-filter feeders, using their nets to trap and concentrate particulate matter in the water. As azamethiphos is liable to adsorb readily onto sediments and suspended particulate matter, it could readily affect these larvae as a stomach poison. Caddis larvae of another family, the Limnephilidae, were unaffected by this treatment, and this must be attributed to the fact that these larvae do not feed on particulate matter, but graze on algae, fungi and minute forms of life on the substrate.

EIGHT

AQUATIC ENVIRONMENTAL EFFECTS

OF INSECTICIDES USED IN

TSETSE FLY CONTROL

INTRODUCTION

For many years, control of tsetse fly in Africa was carried out by a variety of methods based on environmental manipulation, such as bush clearing, game exclusion, habitat destruction by burning etc. The choice of methods was mainly determined by the nature of the habitats characteristic of different species of tsetse, and also by whether the objective of these operations was tsetse control or tsetse eradication.

With the advent of the synthetic insecticide DDT and its allies, increasing emphasis has been on the application of insecticide to the tsetse environment either by means of heavy residual dosages to tsetse-resting sites, or by repeated non-residual applications at lower dosage rates (Jordan, 1974). Initially the insecticides of choice were DDT and dieldrin, the latter being favoured because of its higher toxicity to tsetse. However, it was recognised early that such tsetse control measures had a serious immediate effect on wildlife, mammals, birds, reptiles and fish (Graham, 1964). Over the last 20 years therefore, the preferred insecticide for tsetse control has been the allied organochlorine chemical, endosulphan (Thiodan) (Goebel et al., 1982) selected because of its high lethal effect on tsetse combined with less inimical effect on wildlife (Hocking et al., 1966; Park et al., 1972). That period has also been marked by operational changes; insecticides originally applied by means of ground spraying or fogging equipment, are now applied almost entirely from the air, both by fixed-wing planes and by helicopter.

The increasing scale and progress of tsetse control with endosulphan, and in more recent years by synthetic pyrethroids, in different parts of the African continent has coincided with the increasing worldwide concern with both the short-term and the long-term effect of large-scale application of pesticides in general. This concern has been reflected in the increasing part played by environmental studies in all large-scale operations involving tsetse control with insecticides.

Of considerable significance, in the present context of pesticide impact on aquatic life, has been the fact that so much of the pioneer work on aerial control of tsetse was carried out against the East African game tsetse, *Glossina morsitans*, *G. pallidipes* and *G. swynnertoni*, whose preferred habitats are the vast areas of savannah woodland and thicket, much of it semi-arid in nature, and often remote from water courses and rivers. With the extension of such insecticide-control measures to the West African species of tsetse, namely *Glossina palpalis* and *G. tachinoides*, an entirely different situation was created. These riverine species (*G. palpalis* in particular) are closely associated with dense riverine gallery forest, and with the extensive rivers systems so characteristic of West Africa, and provide a striking contrast to the more arid tsetse environments of East Africa.

Conditions roughly intermediate between these two environmental extremes are provided by another centre of tsetse research and control activity, namely the Okovango Delta of Botswana where a 14000 km^2 inland swamp is infested with tsetse fly, the vector of trypanosomiasis in the economically important cattle population. This area is transected by rivers, interspersed with numerous islands and peninsulas infested with *Glossina morsitans*. Game populations migrate extensively between islands, and the delta supports substantial fish stocks (Davies & Bowles, 1979; Fox & Matthiessen, 1982).

ENVIRONMENTAL STUDIES IN NORTHERN NIGERIA

In northern Nigeria, as in several countries in Africa affected by tsetse, chemical methods of control involving DDT and dieldrin were used for many years. Earlier reports showed that spraying with dieldrin in particular was followed in most cases by the deaths of many wild animals including, among others, insects, fish, birds – especially insectivorous species – and mammals (e.g. bats and squirrels) (Koeman & Pennings, 1970; Koeman *et al.*, 1971. These experiences stimulated more intensified studies on the environmental effects of tsetse control by chemicals, and these studies coincided with significant changes in the method of insecticide application. During the 1960s, insecticides (including DDT and dieldrin) had been applied in a discriminative manner by ground-spraying equipment such as knapsack sprayers, to tsetse resting sites in the dry season at the time when fly populations were most concentrated (Davies, 1971). From 1972 onwards, aerial application by helicopter became increasingly the control method of choice (Challier *et al.*, 1974).

Experiments carried out from 1974 onwards in three different tsetse localities in the Guinea savannah area of Nigeria were designed to study environmental effects of spraying under different conditions (Koeman *et al.*, 1978). These operations were directed against all three tsetse species involved in human or animal trypanosomiasis, namely *Glossina morsitans* in the savannah, and the two riverine species *G. palpalis* and *G. tachinoides*. Spraying consisted of a single application of persistent insecticide to the riverine forests and savannah woodlands, the main targets being the dry-season habitats. In one of the three experimental areas (in the South Guinea zone) only the fringing forest habitats of *G. palpalis* were sprayed.

The experiments were designed to compare the effects of conventional ground spraying (in which the insecticide is selectively applied to such day-time resting places as the lower portion of tree trunks, undersides of horizontal and inclined branches, creepers etc.) and aerial application by helicopter in which the bulk of the spray volume is deposited on the foliage and twigs, with lesser impact on the lower trunks and the ground. The extent to which rivers and other fresh water bodies were exposed to chemical applied by the less discriminative aerial route, is exemplified by the forest research test area where no less than 1400 km of riverine vegetation were affected.

The effects of such treatments on aquatic life were studied in the third sprayed area, where endosulphan was applied to 40 km of riverine forest. Over this whole length, mass mortality of fish was observed. Even $1\frac{1}{2}$ months after spraying, only a few very juvenile fish were collected, and complete recovery was not evident until a year after spraying.

Although no specific observations were made on stream invertebrates in these Nigerian studies, which were mainly concerned with the effects of spraying (with both dieldrin and endosulphan) on bird and mammal populations, the drastic effect on fish populations must certainly have been accompanied by equally disastrous reactions on the part of stream invertebrates. Abundant evidence to support this comes from the investigations which followed, particularly in Burkina Faso (Upper Volta) and in Botswana.

IMPACT OF AERIAL SPRAYING WITH ENDOSULPHAN AND DECAMETHRIN IN BURKINA FASO

In 1975, a World Health Organization (WHO) project started in Burkina Faso in order to study the most efficient way of applying insecticide from the air for control of tsetse fly (Molyneux *et al.*, 1978). The area in question was in the moist savannah in the south, where such environmental factors as the long wet season, denser and less well-defined tsetse habitats, presented much more difficult control problems than in the dry areas of the north. This project started with the full knowledge and recognition of the possible undesirable side effects likely to be produced, and accordingly appropriate environmental studies were initiated at an early stage.

In the dry season of 1976–77 the main technical problem to be overcome was the development of an aerial spraying technique that could be used to apply insecticide rapidly and effectively, particularly as aerosols, to irregular shaped riverine forest habitats of the tsetse, *Glossina palpalis* and *G. tachinoides*, with minimal risk of environmental contamination (Molyneux *et al.*, 1978). The use of the less-manoeuverable fixed-wing plane, unable to follow closely every twist and bend of the river, would have produced a high level of contamination of the river waters; consequently, the helicopter was the obvious choice of aircraft.

Several insecticides were tested, all having biodegradable properties (in contrast to the persistent DDT and dieldrin). Promising among those were endosulphan and the synthetic pyrethroid, decamethrin. The insecticide was applied in the form of aerosol, i.e. the average droplet volume was less than 50 µm (Lee *et al.*, 1978). The site chosen for the insecticide trials was the Kamoe Valley, about 240 km south of Bobo Dioussalou, and typical of the humid savannah region. Spraying operations were carried out in both savannah woodland and in moist riverine forest, the latter habitat being of most significance from the point of view of river contamination. The width of the fringing forest belt ranged from a few metres to a maximum of 400 m, but in most areas was approximately 50 m. The concentration of the main vector species of tsetse in the shaded vegetation bordering the water edge increased the likelihood that effective spraying would unavoidably affect the river itself.

In order to assist in evaluating the impact of spraying on non-targets, a 3.5 km stretch of the Kamoe River was set aside as an untreated control zone. Particular attention had been given in preliminary trials to the dispersal and size of insecticide droplets within the riverine forest and at the river surfaces (Lee *et al.*, 1978), with the normal path of the helicopter close to the river's edge, and usually only 2–3 m above the forest canopy.

Where there was narrow forest around a bend in the watercourse, river contamination was reduced by flying the helicopter almost sideways, with the main rotor disc tilted towards the water so as to direct the downwash away from the river and into the forest (Baldry *et al.*, 1978a).

In 1977 a special investigation was carried out to study the effect of the aerial spraying on non-target fauna (Takken *et al.*, 1978) along the 85 km stretch of fringing forest in the Kamoe Valley. In striking contrast to previously reported drastic environmental effects produced by tsetse control operations using DDT and dieldrin, none of the present series of biodegradable insecticides produced any harmful effect on the bird and mammal populations. The aquatic studies were mainly concerned with fish, but included two large crustaceans, the shrimp *Macrobrachium raridus* and *Caradina africana*. The application of decamethrin (at a dose rate of 0.19 g AI/ha aerosol) led to mass mortality of these two species of freshwater shrimp, at a time when no adverse effects on fish were observed. Most specimens of the affected shrimps were found in a paralysed condition, or dead or dying on the river bank. Somewhat similar, but less extreme effects, were produced by permethrin.

These pyrethroids were also observed to cause high mortality in insects living on or close to the river surface, such as skaters (Gerridae), back-swimmers (Notonectidae) and certain beetles such as *Dytiscus* and *Gyrinus*.

In 1978 further studies were carried out to compare the two main insecticides, decamethrin and endosulphan (Baldry *et al.*, 1981), using oil-based ultra low volume (ULV) formulations of the insecticides, diluted with diesel oil immediately prior to spraying. The object of using the oil-based formulation was to increase the residual properties of deposits on vegetation, producing more effective tsetse control, while at the same time reducing the extent to which contaminant droplets could disperse vertically down through the river water.

Again, in assessing the effect on aquatic non-targets the emphasis was on the fish fauna. On the day following a single application of endosulphan (100 g/ha) large numbers of fish were found dead or dying. Many dead aquatic insects were also observed including Gerridae and Gyrinidae, but dragonfly nymphs, crustaceans and molluscs did not appear to be affected. The effects of a heavier dose (200 g/ha) were very similar, with crustaceans, Odonata and molluscs still unaffected. Of the many species of fish killed, one, *Gnathonemus pictus*, is of particular interest in that of six species of fish whose stomach contents were examined, this one in particular had consumed a greater variety and volume of insect material after spraying, including presumably many dead insects which had fallen onto the water, as well as the real aquatic insects. Examination of stomach contents of species caught before spraying showed abundance of aquatic insects such as mayflies, caddis, dragonflies and larval Diptera. These were also present in the post-spraying stomachs, with the exception of mayfly larvae. This disappearance of mayfly (Ephemeroptera) larvae from fish diet following spraying was also noted in two other fish species examined, *Chrysichthys* and *Synodontis*.

Some light could possibly be thrown on this situation from the results of laboratory tests carried out earlier in a quite different region of Africa, namely in Zimbabwe (Muirhead-Thomson, 1973). Those investigations had shown on the one hand that *Baetis* nymphs (Ephemeroptera) were extremely susceptible to endosulphan, while dragonfly nymphs at the opposite end of the scale were unusually tolerant. It seems quite possible therefore that in the tsetse-control experiments in the Kamoe Valley, the mayfly nymphs would be immediately affected by the spraying operations, and quickly removed by drift from the treated river, thus rendering them no longer available to fish in that area. With regard to the dragonfly nymphs, their ability to survive spraying operations is well in accord with the unusually high tolerance demonstrated in laboratory tests by the several species tested.

One year after the spraying operations in the Kamoe Valley, fish life had apparently returned to normal. With regard to decamethrin, the environmental effects were similar to previous experience, in that while fish populations remained largely unaffected, invertebrates, exemplified by *Cardanina* and *Macrobrachium*, were extensively killed or paralysed.

IMPACT OF ENDOSULPAN AND DELTAMETHRIN ON STREAM MACROINVERTEBRATES IN COMBINED TSETSE/SIMULIUM-CONTROL PROJECTS

The site of these tsetse-fly-control trials comes within the project area of another great disease control programme in West Africa, namely the Onchocerciasis control Programme (OCP), a project which aims to reduce or eliminate the blackfly (*Simulium*) vector of the disease by means of control measures directed against the larval habitats in rivers (see page 189). Although regular weekly applications of larvicides to rivers is the mainstay of that programme, problems were created in some controlled areas by invasions of adult blackflies from outside. Consequently it was decided to carry out trials on the effect of aerial applications of insecticide against the invading adult *Simulium*, which were mainly restricted to riverine forests of the invaded rivers. In view of the experience gained in the tsetse-control trials in that area, it was decided to utilise the 1978 programme to monitor the effect on adult blackflies. The Kamoe itself had been under treatment with *Simulium* larvicides since 1975 and there was no resident larval population in the river. Accordingly, larviciding was discontinued in 1977, allowing larval populations to build up for the trial (Davies *et al.*, 1982)

This provided further opportunities to study the impact of the two insecticides used, endosulphan and deltamethrin on additional important macro-invertebrate organisms, viz. *Simulium* larvae. Deltamethrin applied to 30 km of riverine vegetation at 12.5 g AI/ha eliminated adult blackflies and killed all the *Simulium damnosum* larvae in the river. In contrast, endosulphan applied to 30 km at the rate of 100 g AI/ha was only partially successful against adults, and the residues which were deposited in the river only succeeded in killing young larvae.

IMPACT OF PYRETHROIDS IN TSETSE CONTROL IN

NIGERIA AND THE IVORY COAST

At the same time as the tsetse-control trials were being carried out in the Kamoe River in Burkina Faso in 1977–78, similar trials were also taking place elsewhere in West Africa, particularly in Nigeria and in the Ivory Coast, mainly concerned with the synthetic pyrethroids.

In Nigeria, tsetse-control trials were carried out in Kaduna State with three pyrethroids, permethrin, cypermethrin and decamethrin in a Guinea savannah zone with narrow strips of forest fringing the rivers (Smies *et al.*, 1980). In 1977, sampling of benthos in rivers was carried out simply by dragging 30 cm nets through the river bottom surface, and these observations showed that water beetles and Crustacea (e.g. *Caradina*) were obviously affected. No attempt was made to treat the macroinvertebrate samples quantitatively and the results could only be expressed as presence/absence before and after spraying of a number of representative groups. The samples indicated that Ephemeroptera were most affected, together with Zygoptera and Corixidae, but numbers were small, and it was not possible to differentiate between the effects of the three pyrethroids tested. In 1978 both shrimps and mayfly larvae recurred in rivers which had been sprayed in 1977.

Environmental effects of the two pyrethroids (permethrin and cypermethrin) applied by helicopter were also studied in the rather different context of control measures directed against adult black flies (*Simulium*) in an area included in the OCP in Togo, West Africa, already under larvicide control (Douben, Everts & Koeman, 1985). The heavy dosage used (248–530 g AI/ha of permethrin and 72–198 g AI/ha cypermethrin) to establish barriers against immigrant blackfly adults had the effect of virtually wiping out fish and crustaceans.

As part of the WHO special programme on tsetse control in the Ivory Coast (in the Boufle area northwest of Abidjan) a study was made of the environmental side effects of helicopter spraying with permethrin and deltamethrin on a riverine forest (Everts *et al.*, 1983). In this investigation, the sampling techniques for macroinvertebrates were more refined than in the earlier studies on tsetse control, more in keeping with the well-tried procedures used in the other concurrent vector-control project affecting West African rivers, the OCP.

Benthic drift was sampled by drift nets at 17.30 h, a time of natural increase of drift. Artificial substrates in the form of concrete blocks were also used, these being installed just below the water surface, anchored on a beam. After each spraying, four round scoop nets acting as drift nets were placed in a riffle for at least the night following treatment. Additional random checks were carried out on stones from the river bed.

The gallery forest along the study river was treated five times with deltamethrin (12.5 g/ha) and once with permethrin (40 g/ha). Within a half hour of the deltamethrin spraying there was a sharp increase in the number of Ephemeroptera, Chironomidae and *Simulium* in the drift nets, the drift index (number of organisms per cubic metre) reaching 2800 after the first treatment. The samples from drift, artificial substrates and random collections all indicated serious and immediate effects on Ephemeroptera, particularly in the families Tricorythidae, Caenidae and Heptagenidae; but already after 2 weeks there were signs of recovery. The same comparatively rapid recovery was noted in the Trichoptera from artificial substrate collections after they had been temporarily affected.

The two dominant crustacean species in the river reacted rather differently to the deltamethrin treatment. The larger shrimp *Macobrachium vollenhovenii* (normally hidden during the day) appeared from their shelters, some being found paralysed in the drift nets. Apparently, the adverse effects were only temporary because after 2 days' maintenance in nets under the water, recovery was complete. In contrast, the smaller shrimp, *Caradina africana*, was found dead in large numbers after treatment, and none could be recovered from any collections after the third spraying.

In rocks and substrates which had been denuded of normal fauna as a result of treatment, chironomids were quick to colonise and, 4 weeks after treatment, showed populations far in excess of those in an untreated comparison river. Many other groups of aquatic insects such as dragonfly nymphs, beetles and pond skaters were observed to be immediately affected by treatment, but there was no measurable reduction in their populations.

TSETSE-CONTROL OPERATIONS IN OKAVANGO, BOTSWANA: IMPACT OF ENDOSULPHAN

BACKGROUND AND INVERTEBRATE SAMPLING

The Okavango in northwestern Botswana is a remarkable example of a river system (arising in Angola) which does not finally flow to the sea. It ends up instead in a huge inland freshwater delta covering about 20000 km². It consists of permanently flooded Papyrus swamp in the north, and of seasonally inundated floodplains towards the south. It is also an important wildlife refuge, and the area is fringed by human settlements where stock-raising is an important activity. The importance of the delta lies in the fact that its numerous islands and peninsulas provide the main centre of tsetse fly infestation (*Glossina morsitans*) in Botswana.

From previous experience in other parts of Africa, endosulphan was the insecticide of choice in the tsetse-control operations which began in 1972 (Douthwaite *et al.*, 1981). The need to reduce environmental contamination also dictated that the insecticide be applied, not in high residual dosages, but in small aerosol applications, 6–12 g/ha. In 1974 the UK Ministry of Overseas Development was asked by the Botswana government to advise on all environmental aspects of the tsetse-control operations. As a result of this, a team of scientists carried out investigations from 1975 to 1979 with particular reference to the distribution of endosulphan in the environment, and the impact of spray operations on freshwater fish, aquatic invertebrates and bird life.

Unlike the projects in West Africa described earlier in this chapter, application of insecticide was made by fixed-wing planes (Cessna 310s) and not by helicopter, flying singly or in pairs in echelon. Over a 3–4 month summer period, spraying was carried out at intervals which were short enough to prevent female tsetse from emerging, mating and depositing their first larvae. Spraying was carried out at night when the atmospheric conditions were more stable, with the flight level about 15 m above the tree canopy.

Of the great mass of information provided by the efforts of this team – mainly dealing with fish toxicity (Matthiessen *et al.*, 1982) – two aspects are relevant to the present review, namely, the reactions of aquatic invertebrates and the results of chemical monitoring of endosulphan in the waters of the swamps. Invertebrate studies dealt mainly with the fauna of relatively shallow lagoons rather than the rivers with which they were associated. Because the Okovango Swamp embraces such a variety of water conditions, including rivers and fast-flowing channels, there would be no real advantage in excluding this material on the grounds that it referred to static water rather than running water.

Two approaches to sampling were adopted in assessing the impact of endosulphan spray on freshwater invertebrates in the field (Russell-Smith & Ruckert, 1981). The first involved frequent sampling before and after individual spray events in order to detect short-term effects of the insecticide. The second approach involved sampling a larger number of sites before and after a complete spraying season.

Sampling took two main forms; namely bottle collections of open lagoon water for zooplankton, and collections of all invertebrates associated with clumps of submerged macrophytes. These intensive investigations were unable to demonstrate any dramatic or consistently adverse effect of endosulphan spraying on aquatic invertebrates. The experience was illuminating

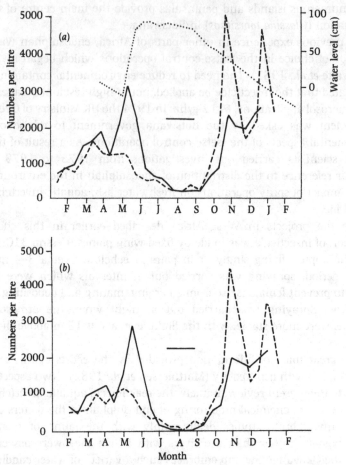

Fig. 8.1. Changes in numbers of zooplankton in an unsprayed lagoon (solid lines) and in a lagoon aerially sprayed with endosulphan (broken lines) in tsetse fly control operations in Okavango, Botswana. (a) Total invertebrates; (b) Rotifera. The horizontal bar represents spray period, and the dotted line illustrates changes in the water level. Plankton density expressed in numbers per litre (after Russell-Smith & Ruckert, 1981).

in one respect in that it showed how difficult it is to evaluate the influence of one factor (insecticide treatment) in the presence of several other factors which may exert a much greater influence on population fluctuations. This is evident in the Okavango Swamp where sharp seasonal changes in population levels of zooplankton (of which Rotifers made up 63–6%) were associated with great increases in water level caused by flooding in the rainy season (Fig. 8.1). The spray season coincided with the period of lowest zooplankton density; by the end of the spray season the water level was already falling and water temperatures rising, and this was followed within the next 2 months by a rapid increase of zooplankton to peak levels.

These experiences also showed that comparison between a sprayed area (lagoon) and an untreated lagoon could prove to be an unreliable basis for evaluation because each lagoon does not react in an identical manner to flooding and changes in water level. In fact, if the charts in Fig. 8.1 were taken at face value, one might conclude that during the 2 months following the end of spraying operations, zooplankton reached very much higher population densities than in the untreated lagoon, and that spraying actually led to an *increase* in zooplankton.

With regard to the invertebrates associated with clumps of vegetation, only chironomid larvae showed significant decreases after spraying, but this effect did not persist for any great length of time. Lower numbers of Trichoptera nymphs were also recorded in the sprayed lagoon through the sampling period, but no such density changes were observed in the case of Ephemeroptera.

The overall conclusion from these intensive field studies is that under the conditions prevailing in the Okavango Swamp, the methods used for applying endosulphan for control of tsetse flies did not have any obvious adverse effect on aquatic invertebrates.

ENDOSULPHAN DISPERSAL IN RELATION TO AERIAL APPLICATIONS RATES

An important part of the investigations in Okavango was to study the pattern of endosulphan dispersal at ground level in various habitats, in relation to the aerial application rates. This was done by arranging a series of aluminium foil squares (20 cm) along a straight line at right angles to the flight path of the aircraft. In one series, 23 squares were arranged in line at 40 m intervals. In a second series, designed to sample several contrasting habitats – including open water and emergent grass swamp – sampling areas were selected at intervals of 3–5 km, again along a line at right angles to the flight path. Endosulphan deposits were assayed by liquid–gas chromatography.

The normal application dosage of endosulphan in the sprayed area was 9.5 g/ha which would theoretically produce a deposit of 38 µg endosulphan per sheet (1 g/ha = 4 µg per sheet). The results obtained however showed very wide variations from this estimated figure. The figures for open water and water in a grass swamp – figures which did not differ significantly from those on open grassland at ground level – were as follows:

	Quantity of endosulphan on sheet (µg)	
	Open water	Swamp
Mean	20.7	17.7
Range	1.4–42.9	3.6–45.9
Variation	116.1	193.4

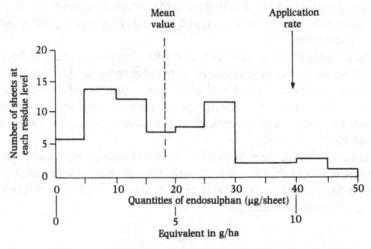

Fig. 8.2. Frequencies of different quantities of endosulphan collected at ground level after an aerial application of 9.5 g/ha (after Douthwaite *et al.*, 1981).

If the distribution of aerially applied droplets had been random, the mean and the variation would be approximately equal. The figures show in fact that the variance is much greater than the mean, and that more samples contain very high and very low values than would be expected. There are thus localised areas of the environment receiving very high and very low concentrations of the insecticide. In extreme examples, due mainly to navigational error on the part of the pilot, aircraft in one instance repeatedly missed one monitoring point, while at the same time passing over another point on three consecutive runs, giving an abnormally high dosage which resulted in fish kills.

These variations are well brought out in Fig. 8.2 (Douthwaite *et al.*, 1981) This shows that only a small proportion of the sheets (approximately 5%) received quantities of endosulphan equivalent to or in excess of the sprayed dosage. Due to a variety of navigational errors, such as overlapping or incorrect spacing, 50% of the habitat was underdosed and 7% overdosed.

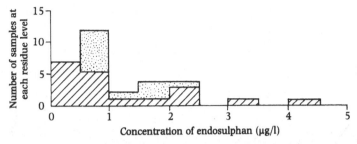

Fig. 8.3. Frequency of different concentrations of endosulphan in water samples taken from a variety of habitats after aerial application at 9.5 g/ha (after Douthwaite *et al.*, 1981). Hatched areas represent 1977 data, dotted areas represent 1978 data.

These figures represent only one facet of a very complex situation as they are based on samples of deposits at ground level taken 1 h after aerial spraying. On this basis the figures may possibly underestimate the true deposit because of some loss of insecticide (perhaps 10%) by evaporation during that hour. In addition, after removal of the sampling plate, the habitat may subsequently be exposed to additional insecticide due to fallout. There are figures to indicate that 9 h after spraying the quantities of endosulphan in shallow water were 40% greater than samples taken 30 min post-spraying.

With regard to the aquatic environment in particular, the Okavango studies concentrated on two aspects of universal interest and importance, namely the range of concentration of insecticide likely to be encountered by aquatic organisms after spraying, and the duration or persistence of such concentrations. With regard to range, theoretical application of around 10 g/ha could produce lethal concentrations in shallow water of 1 μg/l in water 1 m deep, and 10 μg/l in water 10 cm deep (Fox & Matthiessen, 1982). Water samples taken from a wide variety of freshwater habitats 6–9 h after spraying at 9.5 g/ha were collected and analysed (Douthwaite *et al.* 1981). The range of concentrations and the frequency distribution are shown in Fig. 8.3 from which it will be seen that (as in the case of the ground samples) the bulk of the figures show concentrations of endosulphan at the lower end of the scale, from 0 to 2.5 μg/l, much lower than would be expected from the theoretical figures, which at a depth of 20 cm, and spraying at 10 g/l, should show a concentration of 50 μg/l

The results obtained in river water are particularly relevant to the present review. Samples were taken from a depth of 1–2 m, compared to a depth of less than 0.5 m at which the samples from static water bodies of marshes and open pools were taken. Figures revealed a wide range of concentrations from 0.5 to 4.0 μg/l, with a mean of 1.6.

On the question of persistence, endosulphan is one of the least persistent of the organochlorines, with concentrations falling off by 50% within 1–5

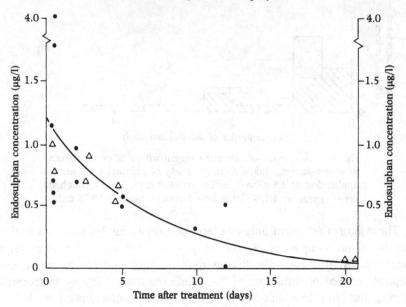

Fig. 8.4. Concentration of endosulphan in samples taken approximately 30 cm below water surface after aerial spraying at 9.5 g/ha in tsetse fly control. Triangles, still water lagoon; filled circles, flowing river. (After Douthwaite *et al.*, 1981.)

days (Gorbach *et al.*, 1971*b*; Goebel *et al.*, 1982). In the Okavango, persistence of endosulphan was tested in a series of experimental tanks or 'ponds' set out in the sprayed area. The results confirmed the rapid decline in concentrations from 9 h after spraying onwards, reaching undetectable levels by the fifth day. In samples taken from deeper water, both in the lagoon and in the main river, endosulphan remained detectable for much longer periods, 10–20 days (Fig. 8.4) (Douthwaite *et al.*, 1981).

NINE

EFFECT ON NON-TARGETS OF LARVICIDES
APPLIED TO RUNNING WATERS FOR
CONTROL OF BLACKFLY
(*SIMULIUM*) LARVAE

INTRODUCTION TO BLACKFLY CONTROL

Blackflies (Simuliidae) are biting flies which are widely distributed in both temperate and tropical regions. In some northern countries such as Canada their main economic importance is as biting and bloodsucking pests of humans and domestic stock. In other regions such as tropical Africa and Central America their main importance is in their role of vectors of human diseases such as onchocerciasis caused by a parasitic filarial worm. A feature common to all species of *Simulium* is their association with running waters which form the larval habitat. According to species and country, these habitats or breeding places may range from quite small trickling streams to very large rivers of Africa such as the Niger, the Zaire, the Volta and the Nile. In many of these rivers and larger streams the highest larval populations tend to be concentrated in the fast-flowing sections of turbulent water such as those associated with rapids and dam spillways.

Shortly after the discovery and rapid developments of DDT in the early 1940s, it was found that *Simulium* larvae are extremely sensitive to this insecticide and that, in some cases, effective larval control was still achieved many miles downstream from the point of application. This opened up entirely new possibilities for *Simulium* control on a large scale by effectively reducing or eradicating larval populations with insecticide. In one of these early and successful control operations DDT was applied at high dosages of 5–10 ppm, or even greater on occasions, which produced massive fish kills and drastic effects on other stream fauna. The discovery of DDT's undesirable residual and bioaccumulation characteristics led to its gradual phasing out during the 1960s. Coincident with this withdrawal, and the search for alternative chemical larvicides, was the increasing concern about the environmental

175

impact of pesticides in general, and about their contaminating effect – direct or indirect – on fresh waters. The present review of *Simulium* larvicides and their impact on non-target macroinvertebrates deals with the post-DDT period.

With mounting information about the capacity of DDT and its metabolites to accumulate in the bodies of fish and other aquatic forms of life, and to persist in the environment for long periods – often for years – after use, attention was turned more and more to possible alternatives for *Simulium* control. There were several possible replacements for DDT, including carbamates and organophosphorus insecticides, but the choice fell increasingly on methoxychlor. Methoxychlor is also a chlorinated hydrocarbon, with low mammalian toxicity, but early evidence indicated that it lacked DDT's dangerous property of accumulating and persisting in the environment. Preliminary trials had demonstrated its effectiveness as a *Simulium* larvicide (Travis, 1949).

ENVIRONMENTAL STUDIES IN NEW YORK STATE ON
THE IMPACT OF METHOXYCHLOR

The possible environment effects of methoxychlor use received particular attention in New York State which had a long history of *Simulium* control with DDT but where the use of DDT was banned in 1964 in lands under control of the Conservation Department and situated in watersheds containing lake trout (Burdick *et al.*, 1968). Experiments were carried out in 1965 and 1966 in hatching ponds stocked with common sunfish and brown bullhead, and these tests showed that in ponds to which methoxychlor had been added no residues were detected in samples taken 36, 63 and 118 days after the final addition. These samples were taken from fish and some other members of the food chain such as crayfish, snails and dragonfly larvae. In contrast, DDT and its metabolites were present in all samples from the DDT-treated ponds.

Having satisfied the main requirement for a likely successor for DDT, it remained to be seen how its *Simulium* larvicidal qualities compared and what effect it had on non-target stream fauna. The established evaluation methods used to study macroinvertebrate populations in streams treated experimentally with methoxychlor in 1965 were drift collections and Surber square foot bottom samplers. The chemical was applied by hand sprayer at concentrations equivalent to those which would have been produced by the single swathe of aerial application. When a fuel-oil formulation was used, a heavy drift of mayfly larvae and stoneflies took place, with a lesser drift of caddis. Blackfly larvae comprised only 9% of the drift. With an emulsifiable concentrate formulation, the *Simulium* population was more severely affected and was reduced by 98.6%.

The importance of these experiments in New York State is that they paved the way for the long series of intensive investigations carried out by the Canadian workers from the early 1970s dealing with the impact of methoxychlor treatment, and especially with a penetrating analysis of drift data. Of particular relevance to the general problem of evaluation was the disclosure in the New York State work that the invertebrates in the drift produced by methoxychlor treatment had not necessarily absorbed a lethal dose of the chemical. Samples of these organisms were kept in trays of clean water for up to $4\frac{1}{2}$ h without any observed mortality. Though not conclusive proof that a portion, or some sections, of the drift could survive methoxychlor treatment, the observation was significant in that up till that time (on the basis of trough experiments with *Simulium* larvae) it had been assumed that drift was an accurate measure of mortality (Jamnback & Means, 1968; Jamnback, 1969).

The workers in New York State were also very conscious of the fact that comparisons between different trials, or between different formulations of the insecticide, could be affected by differences in sampling procedures for assessing invertebrate populations. In trials involving different formulations for example, in the first trial Surber sampling was carried out about 800 m below the point of application, and drift samples also 800 m below. In the second trial bottom and drift samples were taken 200 m below, while in the third trial in which insecticide was aerially applied, drift and bottom samples were taken 1200 m below the application point. In the last trial, the composition of the drift led the authors to conclude that those sampling points were too far below the application point to provide a satisfactory measure of impact.

In selecting the exact sites for bottom fauna Surber samples, the authors, conscious of the fact that population density and species composition in the stream bottom is liable to vary according to depth and distance from shore, adopted the practice of sampling only the central portion of the riffle. They also selected a riffle area of sufficient length and uniformity to allow 18 Surber samples to be taken throughout the season without reworking any portion of the area.

The general conclusion from these preliminary trials was that the immediate impact of methoxychlor on aquatic invertebrates was at least no greater than that with DDT, but with the advantage of non-accumulation and magnification in the environment.

CANADIAN TRIALS IN SMALL STREAMS

WITH NEW LARVICIDES

Canada has a long history of *Simulium* control based on application of larvicide to the streams and large rivers which form the larval habitats. Their pioneer investigations on the impact of DDT on aquatic fauna of running waters have been well documented elsewhere (Chance, 1970; Muirhead-Thomson, 1971; Jamnback, 1976).

Following the discontinuance of DDT, attention was directed towards more selective, less-persistent larvicidal chemicals including methoxychlor and the organophosphorus compound, Abate. In 1969 a comparison between these two insecticides, as well as with Dursban (chlorpyrifos-ethyl, another organophosphorus chemical) was carried out in an area of Quebec, on the north shore of the St Lawrence River, notorious for its high summer populations of *Simulium* (Wallace *et al.*, 1973b). Nine streams were selected, three for each insecticide tested. Four sampling stations were located in each stream at intervals of 402 m and designated *A*, *B*, *C* and *D*. In one of the three streams used for each insecticide, drift nets were positioned at *A*, *B* and *C*, and all streams had a drift net at *C*. Station *A*, standard for all streams, was sited above the application point, while station *B* was situated 137 m below that point. The drift nets, with an aperture roughly 30 cm × 30 cm, were installed in mid-stream, the drift sampling periods being 24 h before and 24 h after treatment.

Surber samples were taken in each stream about 91 m below station *B*, with five samples being evenly spaced across the stream bed. The basic sampling technique specifically designed for *Simulium* larvae took the form of artificial substrates of hollow white polystyrene cones (Wolfe & Peterson, 1958). Counts of larvae were also made on randomly selected rocks. The two organophosphorus larvicides were applied to the streams at the rate of 0.1 ppm for 15 min, while methoxychlor was applied for the same period at the rate of 0.075 ppm.

With regard to the target organisms, *Simulium* larvae, all three insecticides produced a great increase in post-spray drift, indicative of a heavy impact on the population, and this interpretation seemed to be endorsed by the marked reduction on cone and rock counts. However, the Surber sampling revealed the continued presence of larvae. The effect on non-target invertebrates (Ephemeroptera, Plecoptera, Trichoptera and Chironomidae) followed a somewhat similar pattern with very large increases in post-spray drift, indicative of a great reduction in population; however, this was not borne out by the continued presence of all non-targets in the Surber samples.

Again these workers were very conscious of all the pitfalls and possible

178

errors in interpreting the field data. The question arose as to whether the persistence of *Simulium* larvae in Surber samples could be attributed to rapid migration or drift into the sampling area, or whether a proportion of the larvae were those which had re-attached after drift. These trials were carried out at a time when the whole question of invertebrate sampling was being critically reappraised by fresh water biologists elsewhere.

Of particular significance too was the finding that the formulation chemicals alone, viz., fuel oil and heavy aromatic naphtha (HAN), without the active insecticide ingredient, contributed significantly to detachment and drift of both target *Simulium* and non-target fauna, and possibly to subsequent mortality. Questions also arose as to how much the high mortality of drift organisms may have been a function of trauma caused by the drift nets themselves, and also as to the extent to which drift organisms trapped in the net were deprived of any opportunity of regaining and re-attaching to a normal substrate in the stream.

CANADIAN STUDIES ON METHOXYCHLOR AND
INVERTEBRATE DRIFT

In parallel studies by the same Canadian team as above, the drift phenomenon was examined more closely in fast-flowing streams in another area of Quebec (Wallace & Hynes, 1975). In this case the rectangular drift nets were left in place continuously for the 24-h period pre-spray, as previously, but immediately before spraying and for the 3 h immediate post-spray, shorter 15-min periods were used for sampling. After that time a net was left in place for the remaining 21 h of the post-treatment period. Surber samples and cone samples were used in the same manner as previously. The increased frequency of sampling the drift fauna enabled a more precise drift pattern to be established for the different groups of invertebrates. An oil formulation of methoxychlor was applied aerially to one stream and from the ground in another, at the application rate of 0.075 ppm for 15 min.

After the aerial spraying, all the insect groups studied reached peak values in terms of both numbers and weights 75 min after spraying. Plecoptera (stone flies) dominated the drift, while *Simulium* formed the smallest portion. Observations made at this point of maximum drift indicated that the percentage of live specimens of all orders generally increased as time passed until, after 75 min, the number of live specimens drifting had increased appreciably.

In the case of the ground treatment, peak numbers and weight of drifting organisms occurred 60 min after application, and the values were much higher than before, with Plecoptera and Trichoptera (caddis flies) being by

179

Evaluation in pest control projects

far the most numerous drifters. After the peak, drift rapidly declined, falling to very low or 0 levels by 3 h post-spray. As in previous trials, the Surber samples showed that despite the 'catastrophic drift' of all orders of invertebrates studied, non-targets were not eliminated from the treated streams.

Streams in this area of Quebec province were also used for parallel studies on the dispersal pattern of methoxychlor and its downstream transport (Wallace *et al.*, 1973a), but this aspect will be described elsewhere (p. 185).

CANADIAN TRIALS WITH METHOXYCHLOR
IN LARGE RIVERS

THE SASKATCHEWAN

The particular running water bodies used so far for trials and experiments on impact of methoxychlor had all been relatively small streams. Those sited in Quebec for example ranged from roughly 8.5 to 9 m in width, from 0.3 to 0.5 m in depth, and with velocities approximately 0.6–0.9 m/s (Wallace *et al.*, 1973b). Even in the faster-flowing streams the actual discharge was not greater than 2.38 m³/s (Wallace & Hynes, 1975). In striking contrast to these, are the large Canadian rivers which also provide enormous breeding habitats for certain species of *Simulium* which are of economic importance. It was in one of these large rivers, the Saskatchewan, that the earliest observations were made regarding the capacity of DDT, applied as a larvicide, to adsorb onto silt particles and thus be transported long distances downstream (Fredeen *et al.*, 1953; Fredeen, 1962). Since that time, further studies have been made not only in the Saskatchewan River but also in the Athabasca River system, types of flowing water very different from those of the small streams described above. The North Saskatchewan River for example, scene of the investigations now to be described, has an average width of 250 m, an average water velocity of 2.5 km/h, and a discharge varying from 140 m³/s to 494 m³/s (Fredeen, 1975; Fredeen *et al.*, 1975).

After a long history of blackfly control based on larvicide applications of DDT, the phasing out of that insecticide led to its replacement by methoxychlor, which was registered for use as a blackfly larvicide in Canada in 1970. From 1968 onwards a series of trials were carried out in the lower reaches of the southern and northern branches of the Saskatchewan, and in the main river below the confluence, with injections of methoxychlor (Fredeen, 1974). The treated portion of the river contained many rock-filled rapids, ideal breeding sites for *Simulium arcticum*, a biting fly which is a serious threat to livestock in that part of Canada (Fredeen, 1977). Each of the 11 injections carried out from 1968 to 1972 was complete in 15 min at an average

180

concentration of 0.3 ppm, the actual dosage ranging from 0.2 ppm in 7 tests to 0.4 ppm in 1 test. With one exception, an emulsifiable concentrate formulation was used in these trials. In 1968, during a period of stable river discharge, it was possible (through the clear water) to count numbers of *Simulium* larvae in marked colonies on the submerged surfaces of boulders before and after treatment. In the following year, various artificial substrates were introduced to allow quantitative sampling. These took the form of acrylic plates and aluminium mesh plates, each 100 cm² in area, dropped down on rods to within a few inches of the river bed. In addition, vinyl ribbons and short lengths of polypropylene rope were attached to rods near the river surface.

These samplers were set up at various distances from 3 km to 145 km downstream from the injection point. Owing to many variable factors, strict comparison of data from different tests was not possible, but certain consistent features emerged. In these rivers, *Simulium* larvae were much more severely affected by the methoxychlor treatment than were the larvae of such non-target groups as Ephemeroptera, Trichoptera and Chironomidae. In addition, treatment induced drift of *Simulium* larvae much more rapidly than had been the experience in streams. Drift nets demonstrated this rapid 'knock-down' effect, which could also be checked visually in some cases by disappearance of larvae from attachment sites on marked rocks. In one test *Simulium* larvae commenced to release only 6 min after the apparent arrival of the leading edge of the treated mass of water, which at this point – 6.4 km downstream from the injection site – took about 30 min to pass. At the high application rate of approximately 0.3 to 0.4 ppm for 15 min, treatment removed more than 95% of *Simulium arcticum* larvae from the rapids for distances at least 64 km downstream in 2 tests. In another test, 46% of the *Simulium* larvae disappeared from a site 142 km downstream from the nearest injection site.

If the speed of recolonisation be taken as the ultimate measure of impact on larval populations, the larvae of Trichoptera and Chironomidae rapidly recolonised treated areas, while *Simulium* and Plecoptera were slow to recolonise. In the case of chironomid larvae, this lesser impact was associated with the fact that they abound in the sandy bottom of the river where they escape the main flow of insecticide. With regard to *Simulium* larvae, it is suggested that their vulnerability is not only associated with the nature of their aggregation sites on rocky surfaces exposed to the main larvicide stream, but also with the fact that their filter-feeding habits result in the ingestion of fine particles of silt onto which methoxychlor has been adsorbed.

In 1973 a new experimental injection site selected in the North Saskatchewan River allowed the opportunity to study the effects of a single injection of methoxychlor in an uninterrupted section 161 km long. In this case the emulsifiable concentrate formula was applied for a shorter period, 7.5 min, but at a higher concentration of 0.6 ppm. This treatment had a drastic effect on the target organisms (larvae of *Simulium arcticum*) 100% of larvae being eliminated at distances between 40 and 80 km downstream. At the sampling site 161 km downstream, the injection still eliminated 66% of the older instar larvae and 96% of the more sensitive early instar stages less than 1 mm long.

Again, if rate of recolonisation of various artificial substrates be used as the ultimate measure of methoxychlor impact, it appears that despite the immediate dramatic effect, impact was comparatively short lived. All sections of the fauna showed a remarkably rapid return to normality, some more quickly than others; Chironomidae and Ephemeroptera within 1–2 weeks, Trichoptera within 1–3 weeks, *Simulium* 2–4 weeks and Plecoptera 4–5 weeks.

In fact, the re-establishment of those populations during the 10 weeks following treatment was not just to normality, but actually to higher levels than pre-treatment figures in all the non-target groups. This did not occur in the case of *Simulium* larvae. An important part of these investigations was concerned with analysing the pattern of downstream flow of injected methoxychlor (Fredeen *et al.*, 1975). This will be reviewed more fittingly elsewhere.

THE ATHABASCA PROJECT

Introduction While these studies on the Saskatchewan River were being carried out in 1968–72, important developments were taking place in the adjoining part of the neighbouring Canadian State of Alberta. These developments eventually led to one of the most intensive, multidisciplinary studies ever carried out on pesticide evaluation in a large river, (Haufe & Croome, 1980), matched only by another great *Simulium*-control project initiated at about the same time, namely the Onchocerciasis Control Programme in the Upper Volta Basin of West Africa.

A biting fly problem similar to that in Saskatchewan had long been recognised in North Alberta, with periodic heavy loss to livestock. Serious outbreaks in 1962 and again in 1971–72, coming at a time of rapid expansion of the livestock industry, emphasised the need for serious study on the feasibility of control. Surveys carried out were soon able to show that the Athabasca River was the principal breeding habitat of the biting fly responsible, *Simulium arcticum*, and that it was from this immensely productive breeding ground that adult black fly were able to travel long distances of 150 km or more, and infest areas remote from the river.

The Athabasca is a large, turbulent swift-flowing river, draining northeast from the Rocky Mountains into Lake Athabasca. With a total course of 1150 km it is one of the largest rivers in Canada. The mean discharge at the town of Athabasca in 1974 was 572 m³/s, but rose in the same year to a maximum of 2520 m³/s.

Studies in the programme were focussed on the downstream section of the river between a point 65 km above the town of Athabasca and the river delta at Lake Athabasca (Haufe, 1980; Haufe & Croome, 1980).

A feasibility study was carried out in 1973 to define the extent of the breeding areas, to determine the prospects of chemical control by means of *Simulium* larvicides, and to estimate the range of scientific studies necessary for a complete evaluation of chemical impact on both the target fauna (*Simulium* larvae) and non-target biota. The full research programme was initiated in 1974, and the 4-year study was completed with a post-treatment base line investigation in 1977. Methoxychlor was selected as the most appropriate larvicide, the dosage rate being determined as 0.3 ppm applied over 7.5 min.

The great importance of the Athabasca Project against the broad background of river contamination by pesticides in general was its multi-disciplinary approach in which all aspects of the problem (hydrometric, chemical monitoring, and impact on aquatic ecology) were tackled by separate teams of specialists. Particular emphasis was laid on the hydrological studies as it was felt that a full knowledge of the water movement across all sections of the river, and under all conditions of normal flow and of flood with regard to turbulence and mixing, was an essential requirement, both for the interpretation of ecological data and for an understanding of the pattern of chemical transport (Charnetski, Depner & Beltaos, 1980; Haufe, 1980).

A striking feature of the feasibility approach was that the broad problem of evaluating larvicide impact on non-target invertebrates was tackled by three separate teams of scientists, each team bringing a different approach to such basic problems as the choice and validity of sampling methods, the methodology of investigating invertebrate drift, and the interpretation of sampling and monitoring data. The investigation started with the full knowledge that 'there are still no definite criteria or standards, even in general principles, that can be applied in assessing perturbations in ecosystems', and that 'there is very limited knowledge as background for the evaluation of environment impact of pesticides in a large river system' (Haufe, 1980).

Consequently, the scientists on the Athabasca project approached this new situation in an open-minded exploratory fashion. Two of the three teams dealing with the vital question of the effect of chemical treatment on the invertebrate population of the river, designed their programme in close association with the hydrometric survey and the chemical-monitoring programme. The third team worked quite independently. From this situation arose one of the quite unpredictable outcomes of the whole multidisciplinary project, in that the final conclusions of the third team were at variance with those of the other two teams on the key question of whether or not methoxychlor treatment of the Athabasca River was environmentally acceptable (Haufe, 1980).

In order to understand the differences in approach, methodology and data interpretation which produced a conflict of conclusions on a major issue, it is essential to examine these studies in some detail. Before doing so however, it should be pointed out that the main practical objective of this project (namely, the effective control of the target larvae of the pest blackfly *Simulium arcticum*) was achieved, and opinions on that point did not differ significantly (Depner, Charnetski & Haufe, 1980).

First trials with methoxychlor (0.3 ppm for 15 min) The main sampling method used exclusively for *Simulium* larvae were plastic cones providing artificial substrates, suspended on lines at 0.3 and 1.5 m below the surface of the water and anchored in place; cones being retrieved and replaced every 7 days. In 1974 the methoxychlor in EC formulation was applied from a bridge over the river at the rate of 0.3 ppm for 15 min, this period being reduced to 7.5 min in 1975 and 1976. A double treatment was carried out in 1976, while in 1977 no treatment was carried out in order to assess the effect of larviciding.

To summarise a great deal of carefully recorded data on trials carried out in different parts of the river, and at different current velocities, it was shown that at current rates in excess of 560 m³/sec, a 7.5 min injection of 0.3 ppm of methoxychlor gave effective control of *Simulium arcticum* larvae for a distance of at least 160 km downstream. If river-discharge rates are below this point, effective control is reduced to 120 km, and in practice would have to be carried out about every 100 km of river.

To return to the major environmental issue concerning impact of methoxychlor treatment on drift and standing crop of aquatic invertebrates, the main features of the methodology and conclusion of the independent team are as follows (Flannagan *et al*, 1979; Flannagan, Townsend & de March, 1980*a*). In June 1974, the injection studied was the one noted above at the rate of 0.3 ppm, for 15 min. Sampling stations were located at four points: a control station upstream of the application point; a station 200 m downstream; a station 67 km downstream, and a fourth station 400 km downstream. Drift samples were carried out every 4 h, starting 24 h before treatment. The drift nets were of two kinds, standard nets and the 'bomb' net designed to reduce the inflow of turbulent water into the net orifice (Burton & Flannagan, 1976; (see also p. 124)), the latter being considered best suited for fast-flowing water. Artificial substrates in the form of 'barbecue' baskets were also used. An additional technique tested, but subsequently rejected, was the modified Eckman grab, which is standard equipment used by freshwater biologists.

Effects on non-targets of control of blackfly larvae

The conclusion from the three sampling methods was that the methoxy-chlor treatment was followed by catastrophic drift of non-target invertebrates, followed by large decreases in both drift and standing crop for distances up to 400 km downstream. Unlike the experiences reported in the Saskatchewan River, there was no rapid recovery, and very little recolonisation 4 weeks after treatment.

Second trials with methoxychlor (0.3 ppm for 7.5 min) In the following year (1975) the methoxychlor treatment evaluated was 0.3 ppm for 7.5 min; the effects of this dosage at half the 1974 rate were essentially the same, with long term effects noted in Ephemeroptera, Plecoptera and Trichoptera. In fact the catastrophic drift of non-targets was still evident much further downstream than in the case of *Simulium* larvae exposed to the same treatment, suggesting that possibly the methoxychlor was actually more specific for some of the non-targets than for the target *Simulium* (Flannagan *et al.*, 1980*b*).

In the following year, 1976, the whole question of methoxychlor impact on invertebrate drift was re-examined by a separate team, using rather different methods of approach (Haufe, Depner & Charnetski, 1980*a*; Haufe, Depner & Kozub, 1980*b*). Two methoxychlor treatments applied to different reaches of the river with an interval of 6 days, for control of blackflies, were monitored in succession. The continuous monitoring was carried out over a 5 day period which bracketed the expected time of arrival of the methoxychlor pulse from upstream. This close scheduling of sampling period according to hydrological estimates of the expected time of arrival of the pulse of larvicide, and according to the time taken for this pulse to pass the sampling point, is a significant feature of this team's approach. The range of concentration profiles of methoxychlor to which stream fauna were exposed in these treatments is well illustrated in Fig. 9.1. At 18 km downstream from the point of injection the pulse of chemical shows little change from that at application, the pulse lasting about 1 h with the bulk of the chemical passing the sampling point within about 30 min (Fig. 9.1(*a*)). But at 77 km downstream, the duration of the pulse had increased to 10–12 h (Fig. 9.1(*b*)).

Drift samplers were set up at two different parts of the cross-section of the river, and at each point continuous samples were collected at two levels, viz., about 50 cm below the surface, and at mid-depth. Each drift sample removed from the net was sorted out according to whether the fauna were dead or showing signs of morbidity on the one hand, or appeared to be live and healthy on the other. The latter were retained in clean water 3 h in order to further confirm their status. Making allowance for the possibility of trauma sustained by compaction and compression of organisms with drifting debris in the samples, it was possible by this procedure to demonstrate distinct differences in reaction to the insecticide pulse on the part of the 46 different

185

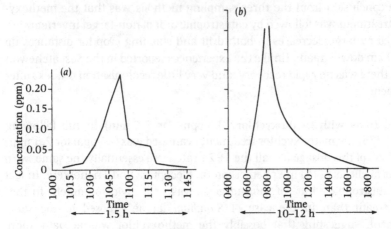

Fig. 9.1. Concentrations of methoxychlor in the water of Athabasca River, Canada, during passage of the pulse at two sampling stations. (*a*) 18 km downstream from injection site; time of passage of pulse 1½ h. (*b*) 77 km downstream from injection site; Time of passage of pulse 10–12 h (after Haufe *et al.*, 1980*a*).

genera of macroinvertebrates represented in the catch. For most organisms the increase in drift pattern was transitory and coincided with the arrival of the front of the larvicide pulse. It was considered that this activity was still well within the range of natural variations due to normal diel periodicities, current-related disturbances and flood. Casualty rates attributable to the pulse were less than 40% for all taxa identified in the downstream drift. Of 49 genera, 37 exhibited no casualties and only 9 genera disclosed consistent sensitivity. These were *Hydropsyche* and *Cheumatopsyche* (Trichoptera), *Baetis*, *Heptagenia*, *Ephemera* and *Rhithrogena* (Ephemeroptera) and *Isogenus*, *Isoperla* and *Hastaperla* (Plecoptera)

Immediate and long-term effects of methoxychlor treatment In close conjunction with the above drift studies, and as part of the multidisciplinary evaluation programme, studies were carried out over the same period to determine the immediate and long-term effects of river treatment on the resident populations of invertebrates (Depner *et al.*, 1980). The need for this separate investigation was indicated by the fact that it had not yet been possible to establish the exact relationship between drift invertebrates and resident populations. Many of the river invertebrates were unrepresented in drift samples, and all indications were that drift only represented a very small proportion of the total standing crop.

The development of appropriate sampling methods was initiated in 1973, the year before treatment started, and in view of the scarcity of previous knowledge of large fast rivers, was exploratory in nature. The technique for bottom sampling was simple and basic, involving disturbance of standard area of bottom and collecting dislodged fauna in a net held downstream – on much the same lines as the 'kicking' technique (see page 122). Over a period of 4 years, these methods yielded a total of 108 genera of non-target invertebrates. Sampling covered the treated years (1974–76) and extended into the non-treatment year (1977). The immense amount of work involved was well rewarded in that the results proved to be of the greatest possible significance, not only in terms of the Athabasca situation but viewed against the general problem of river contamination.

Sampling in the untreated portion of the Athabasca River upstream showed that the diversity (i.e. the number of kinds of organisms) remained relatively stable but, that at the downstream portion of the river (at all stations from 5–240 km), diversity *increased* within each year of the test. The indications were that as populations of some of the more predatory or more sensitive organisms were lowered, an opportunity was provided for some of the less numerous fauna to increase in numbers in the last year of treatment.

In the overall evaluation of all the work carried out on the impact of methoxychlor treatment on non-target invertebrates, several important features are stressed (Haufe, 1980). It was considered that much of the increased drift was associated with the arrival of the leading edge of the downstream slug of insecticide. This increase was particularly marked in the eight more sensitive genera listed above; but at no time could the extent of drift be regarded as 'catastrophic', i.e. sufficient to deplete seriously the standing crop of invertebrates or their ability to recover and regenerate. Much of the increased drift is associated with the long-established characteristic of methoxychlor to cause hyperexcitability, activating the fauna to detach from their substrates and enter the main stream.

Another important point which emerged was the need to interpret sampling data in the light of 'phenology' i.e. the seasonal pattern of invertebrate change from the aquatic to the aerial stage. The particular timing of the methoxychlor treatment, aimed at achieving the maximum possible effect on the target *Simulium* larvae, is also one marked by peak emergence of some non-targets such as Plecoptera (stoneflies). This mass emergence itself plays a major part in the temporary depletion of aquatic populations.

A further point of general importance was the initial degree of mixing of the injected chemical. Simultaneous injection of the larvicide into the river at several points across the river (from power boats) as was done in 1975 and 1976 produced a much quicker mixing of insecticide than did the single midstream injection (from a bridge) of 1974. Insufficient mixing is liable to produce zones or pockets of high larvicide concentration which can have a more serious effect on non-targets; the same effect may be produced by formulations which do not mix effectively and thus produce layered belts of very high concentration at or near the surface. It is suggested that even more rapid and effective mixing would be achieved not only by simultaneous lateral injection but also by simultaneous injection at varying depth in the river, and not just at the surface.

In planning future strategies of *Simulium* control on the Athabasca, all these points would be taken into account. For the purpose of monitoring the effects of the treatment on non-targets it was proposed to use the six sensitive genera as indicators of environmental hazard, and to concentrate on the modified drift sampler as the capture technique most appropriate for such a large, fast-flowing river.

The question of formulation mentioned above in connection with enhanced mixing of the chemical, is also a major consideration in another continuing line of research. That is, to find a formulation which will be selective for the target *Simulium* larvae, but have minimal effect on non-targets. So far most attention has been given to the development of particulate formulations, well-suited to the filter-feeding habits of *Simulium* larvae (Helson & West, 1978; Sebastien & Lockhart, 1981). Progress in that direction will play an important part in determining environmentally acceptable strategies for *Simulium* control in the future.

So far the indications from laboratory studies carried out in connection with the Athabasca project (Sebastien & Lockhart, 1981) do confirm that particulate formulations differ significantly in their effect on some stream fauna. A comparison between a particulate formulation and the emulsifiable formulation used in practical control showed that they were equally toxic to *Simulium* larvae, but with the significant difference that the particulate formulation is more slow acting and that by the time larvae become detached they have already absorbed a lethal dose. This may possibly be an advantage over the conventional EC formulation which produced a rapidly acting effect so that many larvae are induced to enter the drift before they have absorbed a lethal dose, and consequently may survive to re-attach later. With regard to non-target reaction, studies were carried out on stonefly nymphs (*Pteronarcys dorsata*) and on chironomid larvae in a recirculating flow-through system. With both these indicator species, the EC formulation produced a more rapid mortality than the particulate one, most stonefly nymphs being dead within an hour of exposure; but with the particulate formulation, signs of morbidity did not appear until 10 h after exposure. After a 24-h exposure to 0.3 mg/l, all the nymphs exposed to the EC formulation were moribund compared to only 25% with the particulate formulation. The EC formulation also produced a 98% mortality in chironomid larvae as compared with 21% when exposed to the same concentration of methoxychlor in particulate form. The particulate form also produced a less rapid and less complete abandonment of burrows by chironomid larvae.

Effects on non-targets of control of blackfly larvae

The Athabasca investigations covered in this review mark the end of a phase of intensive field studies, and there have been no significant publications on this project since the Technical Report of 1980 (W. O. Haufe, 1985, personal communication). The lessons from the project will continue to have far-reaching effects as they have raised serious questions of methodology to be settled in connection with any future studies of ecosystems in large rivers. While methoxychlor continues to be the pesticide of choice for blackfly control in the large rivers of the northern regions of Canada, there are clear indications that research will have to come up with more satisfactory methods of evaluating the biological impact, particularly with regard to the acceptability of the continuing practice of river treatment with toxic chemical larvicides (W. O. Haufe, personal communitation).

THE ONCHOCERCIASIS CONTROL PROGRAMME

IN THE VOLTA RIVER BASIN, WEST AFRICA

INTRODUCTION

The Onchocerciasis Control Programme (OCP) is a collaborative effort between the World Health Organization (WHO), other international agencies such as the Food & Agricultural Organization (FAO) and the World Bank, and the seven West African countries involved, including Burkina Faso. The object of this programme is to control the human disease, onchocerciasis, caused by a filarial worm parasite and transmitted by the bite of the black fly, *Simulium damnosum* (sens. lat.).[1] In areas of high biting density by black flies and continuous exposure of humans to reinfection, there is increasing likelihood of the parasite invading the eyes and producing the characteristic 'river blindness'. The vast network of the Volta River – the main target of the programme – its numerous tributaries, and other river systems in this part of West Africa (Fig. 9.2) provide ideal breeding grounds for the vector blackfly (Davies *et al.*, 1978). As long ago as the mid-1940s it was shown that *Simulium* larvae, which live attached to substrates in the swift-flowing parts of the rivers, are highly susceptible to DDT and allied synthetic insecticides. It is this vulnerable stage of the blackfly's life cycle which is the spearpoint of attack in the OCP, the main larvicide of choice in the project area being, not DDT (rejected because of its undesirable environmental effects) but an organophosphorus chemical temephos (Abate). Since the OCP officially started in 1974, regular applications of Abate to the Volta River system have been made on an ever increasing scale. This has been carried out mainly by aerial application onto the river surface, both by fixed-wing planes and by

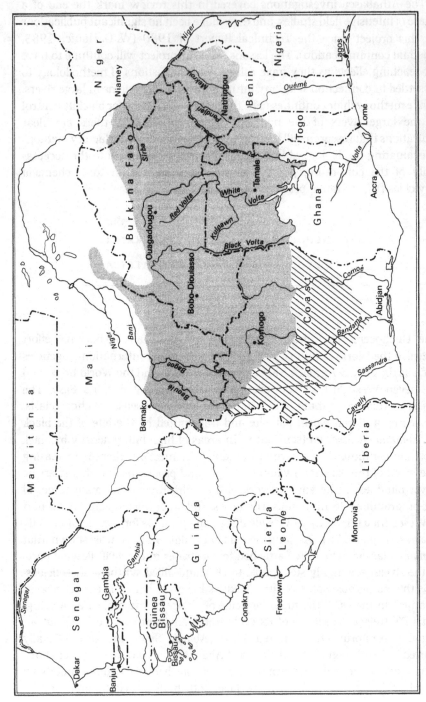

Fig. 9.2. Map of West Africa showing extent of the programme area of the Onchocerciasis Control Programme (OCP) and locations of the main *Simulium*-breeding rivers.

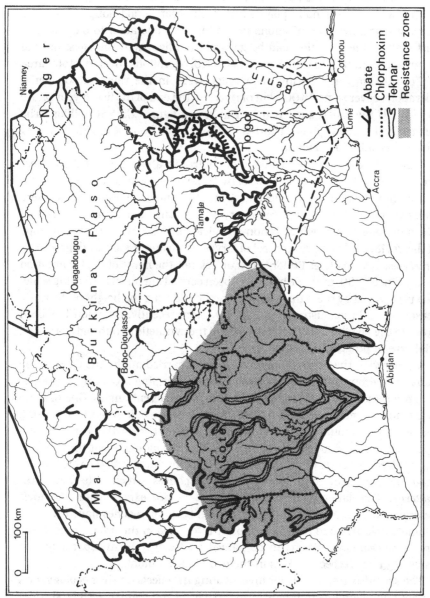

Fig. 9.3. OCP programme area in 1983 under control by the larvicide Abate, chlorphoxim and Bti (*Bacillus thuringiensis israeliensis*) (from OCP, *Progress Report*, 1983).

helicopter, the latter being particularly suitable for pinpointing particularly important sections of the rivers fringed by heavy forest (Philippon *et al.*, 1976*a,b*; Quillevere *et al.*, 1976*a,b*; Stiles & Quelennec, 1977).

The OCP was originally planned for a 20 year period, long enough for existing worm parasite infections in the human population to die out. The vast scale of the operation can be gauged from figures in the latest annual progress report. In 1983 the *Simulium* control operations covered a total area of 764000 km², and during an average week in the rainy season aerial operations cover some 18000 km of river. By the end of the first 6 year phase (1974–79) the programme had cost in round figures US \$56 million. The budget for the second phase, 1980–85 is US \$126 million (OCP, 1985*b*; World Health Organization, 1985*a*).

While Abate is still the mainstay of control in about 80% of the project area, there has been increasing use of two other control measures since 1980. One of these is an allied chemical larvicide chlorphoxim, used in areas where *Simulium* larvae have proved refractory or 'resistant' to treatment by Abate. The other alternative, which is non-chemical and therefore outside the scope of this review, is biological control of larvae by means of *Bacillus thuringiensis israeliensis* referred to as Bti for short, or Teknar, the name of the formulation in use. Since 1980, trials with Bti have been carried out on an increasing scale. The part played by these three control methods during the height of the rainy season in 1983 is illustrated in Fig. 9.3. More recently, attention has been given to the possibilities of permethrin and the synthetic pyrethroids, and field trials are in progress.

In reviewing the immense amount of work carried out by the OCP on the aquatic environmental implications of their continuing programme, evaluation problems have to be approached from two different angles. One of these is the aquatic monitoring of treated rivers in the whole project area using mainly conventional sampling methods. The other is the experimental approach designed to provide more accurate information about the comparative tolerance levels of different groups and species of invertebrates to the larvicide treatment, and to define the responses which determine drift patterns. This latter approach utilises the *in situ* experimental channels described in Chapter 4 (page 56). Although both approaches have the same ultimate objective and have been closely integrated in the programme, there are many valuable lessons to be gained by treating them separately and assessing their respective roles in the overall evaluation.

The scientists responsible for investigating the effects on river ecology, and on aquatic macroinvertebrates in particular, of the regular weekly applications of larvicide for *Simulium* control in the Volta River Basin, were faced with a truly formidable task. The numerous rivers covered a wide range of size and topography, and flowed through widely different climatic and vegetation zones. In the dry season, the water volume in the rivers decreased steadily;

by the end of the dry season short, shallow stretches of fast-flowing water tended to alternate with wide, deep river pools of slow-moving water. In the rainy season, the rivers in flood fill the river bed from bank to bank, covering previous sampling sites and in many cases rendering large sections of the river unapproachable from the land. As the control measures directed against *Simulium* larvae have to be carried out at regular weekly intervals throughout the year, sampling methods have to be devised to cope with all these extremes of conditions.

When the project started in 1974 knowledge of the natural aquatic fauna of West African rivers was very fragmentary and there was an almost complete absence of continuous studies on river ecology covering all seasons of the year. In addition to this difficulty, the scientists in the OCP were not always in a position to carry out adequate pre-treatment studies on rivers in the project area, and in some cases rivers had already been regularly treated with chemical before any aquatic studies could be started.

Despite all these obstacles, combined with the numerous physical difficulties and hazards encountered by survey teams in a most exacting tropical environment, it has been possible in the last 10 years to build up a truly impressive amount of new information about the environmental effects of country wide river contamination by a chemical pesticide (Wallace & Hynes, 1981).

AQUATIC STUDIES ON THE IMPACT OF THE LARVICIDE TEMEPHOS (ABATE)

The aquatic studies on macroinvertebrates fall naturally into two main objectives. Firstly, the short- and long-term effects of regular control treatments on the target organism itself, *Simulium damnosum* s.l., and secondly, the overall effect of larvicide treatment on the many different orders, genera and species which constitute the non-target fauna.

For the purpose of this review, it would be convenient to consider first the general strategy used in evaluating the effects of control measures on aquatic macroinvertebrates in general, and then to examine the additional study methods designed specifically for *Simulium* larvae.

For sampling areas of river bed, the Surber sampler was used throughout the programme. In the monitoring programme the standard area sampled was 15 cm × 15 cm, a metal mesh being placed across the upstream opening to reduce the possibility of drift organisms entering the sampler (Leveque *et al.*, 1977). As the sampler cannot be used in water deeper than about half an arm's length, its use in general was restricted to the dry season. Sampling carried out on two different types of river bed surface, flat-surfaced rock and gravelly bed, showed wide differences in macroinvertebrate composition in the two sites, the latter being richer in numbers and species (Samman &

Thomas 1978*a*). Extensive sampling showed that there was no uniformity in distribution of fauna in river beds a factor of great importance in the interpretation of sampling data.

Later in the programme it was found possible to extend the use of the Surber sampler to the rainy season, using a helicopter to allow team members to collect samples from partly submerged rock surfaces in mid-stream rapids and other sites which would otherwise have been completely inaccessible in such large rivers in flood.

Sampling by drift net proved to be the most consistently reliable technique in the programme as it could be used in a standard manner irrespective of the season of the year. The mouth of the net measured 25 cm × 25 cm and the mesh aperture was 300 μm. The length of the net was extended to 2 m in order to increase the filtering surface (Leveque *et al.*, 1977). The nets were arranged in several positions across the river, and located so that the top of the frame was 2 cm below the surface. Drift samples were carried out at two periods; a day sample approximately $1\frac{1}{2}$ h before sunset in which 3 nets were used for 30 min, (see Fig. 6.2) and a night sample taken approximately $1\frac{1}{2}$ h after sunset, and designed to sample the natural increase in drift at that period. At that period, 6 nettings of 3 min duration was the method adopted.

Artificial substrates, mainly designed to sample organisms not normally available in the drift, were also used in the form of small concrete blocks installed in the river bed and allowed to colonise over a period of 3 months (Leveque *et al.*, 1977). This method has the advantage of being usable at all seasons of the year. In some trials an additional floating artificial substrate was used composed of strips of plastic bound together to form a ball which was submerged in the current and suspended in position from a cable (Dejoux *et al.*, 1982). Normally, such new artificial substrates are quickly colonised by stream fauna, and the main function of this particular design was to measure the rate of recovery of stream fauna following treatment with larvicide, which in this particular case was chlorphoxim. The substrates were set in position on the day of treatment and kept in position for 15 days. Like other artificial substrates, this type was utilised by some groups of inverte-brates but not others, the main colonists being simuliids, hydropsychids and chironomids (*Chironomus*, *Orthocladius* and *Tanytarsus*), as well as oligochaetes.

The degree of colonisation of such substrates was affected by the current velocity in which they were immersed, the most favourable range for most species in uncontaminated water being 75–125 cm/s. Under the effect of chlorphoxim, this colonisation pattern changed, with most colonisation taking place in the 25–50 cm/s range, falling sharply at current velocities in excess of 1 m/s.

The use of these floating substrates also revealed distinct differences between Abate and chlorphoxim as regards colonisation after treatment.

194

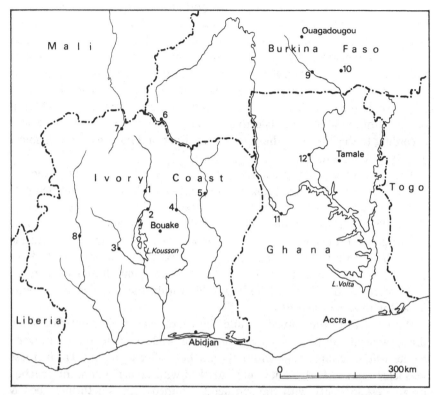

Fig. 9.4. OCP programme area showing location of monitoring
stations (after Leveque *et al.*, 1977).

Under rainy season conditions the overall reduction in colonisation with
Abate was 50–55%, as compared to up to 85% with chlorphoxim (Dejoux
et al., 1982). This is in accord with differences observed between the effects
of these two major *Simulium* larvicides using different study methods (viz.,
drift nets) chlorphoxim consistently producing a much higher rate of detachment and drift than Abate.

In the overall monitoring programme of the project, the distribution of
regular sampling stations is shown in Fig. 9.4, these being selected according
to a wide range of conditions and spray history. Some of these stations were
in untreated rivers not yet affected by the expanding programme, some in
untreated parts of rivers already under regular treatment, and others in rivers
already under treatment. In many of these stations it was still possible
therefore to obtain substantial baseline data about invertebrate numbers and
distribution under natural uncontaminated conditions.

With regard to the target fauna, larvae of *Simulium damnosum* s.l., while
various artificial substrates have been tried out and found useful in some

195

trials, in the project area as a whole the most practical measure of population change in relation to treatment has been the regular inspection of known natural aggregation sites of *Simulium* larvae. In such places visual inspection can best demonstrate drastic reductions or elimination of these populations following application of larvicide, at all seasons of the year. In the high floods of the rainy season, otherwise inaccessible breeding sites can still be readily inspected by means of helicopter. These regular inspections also provide vital practical information about effective carriage downstream of each treatment according to the season of the year, type of river and nature of the chemical and its formulation.

The continuous survey carried out over several years in the extensive project area of the OCP spanning 7 countries has produced a massive amount of data which only a recent 500 + page thesis can do full justice to (Elouard, 1984). The analysis and interpretation of all this information collected under the auspices of the WHO has also been made the responsibility of an outside body (Salford University, 1981) and has proved to be a formidable task. In view of the wide range of experiences reported which vary from river to river, and according to season, it is only possible within the confines of this book to deal with salient features.

With regard to the immediate and short-term effects, drift net data show that practically all treatments with larvicide produce an increase in invertebrate drift, reaching a peak within the first hour after treatment. The increase was least marked with Abate, more marked with chlorphoxim and reached its greatest intensity with the synthetic pyrethroid, decamethrin (Gibon & Troubat, 1980).

Using the ratio (drift index after treatment)/(normal pre-spray index), figures obtained were as follows:

Abate (0.05 ppm)	43
Chlorphoxim (0.025 ppm) [2 formulations]	152–86
Decamethrin (0.007 ppm)	511

Drift samples showed that some organisms are more affected than others. Baetids and hydropsychids (particularly young instars) often form the bulk of the drift samples, and with some species this increase in drift is accompanied by reduction in standing crop as judged by Surber samplers or artificial substrates. For example, in one river investigated (the Oti river near its point of entry into the Volta Lake in northern Ghana) Surber samples showed that baetids and leptophlebids among the mayflies were seriously affected by Abate treatment, while caenids appeared to be unaffected (Samman & Thomas, 1978b). Of particular significance in that investigation was the observation that the most seriously affected organisms were the dragonfly (Odonata) nymphs of Gomphidae and Libellulidae, as well as dytiscid beetles. These are all-important predatory organisms, present in relatively low densities and rarely represented in artificial substrates or in drift net samples.

196

In the case of chlorphoxim, as with Abate, the great increase in drift of baetids is accompanied by a fall in standing crop as judged by Surber samples (Gibon & Troubat, 1980) but there is in addition a marked effect on hydropsychid (caddis fly) populations in the rapids which form their habitat (Dejoux *et al.*, 1982). Among the chironomids there is a sharp difference in reaction between the Tanytarsini which are adversely affected and the Orthocladiinae which actually show an increase after chlorphoxim treatment, possibly related to the destruction of their normal predators.

Despite the great amount of information from field surveillance data, and the general consensus on the immediate effects of the various larvicide treatments, further analysis of invertebrate reaction to the chemical contaminant would have been extremely difficult without recourse to controlled experimental methods. The main technique used in this project, viz., channels installed *in situ* in the natural river habitat (page 56) has indeed made it possible to make accurate comparisons between different invertebrate groups, different insecticides and different treatments in a way that would not have been possible from field sampling and surveillance alone. The results of this experimental work are reviewed later.

ASSESSMENT OF FIELD MONITORING DATA

The question of possible long term effects of regular applications of larvicide to the river systems in the OCP project area has long been a matter of great concern to the biologists involved as well as to the WHO, the international agency with most responsibility for environmental aspects of the programme (WHO, 1984, 1985 *a,b,c*).

The early observations indicated that although larvicide treatment produced sharp increases in the number of drift organisms there was little evidence that this drift was 'catastrophic' in the sense that it seriously depleted the standing crop of non-target invertebrates. Even with the more sensitive groups, baetids for example, any reduction in bottom fauna following increased drift was temporary and was quickly restored. However, the very fact that treatment had a differential effect on different groups, and that some were clearly more sensitive to the larvicide treatment than others, suggested that changes in community structure might be produced which could eventually have a serious effect on the river ecology as a whole.

After several years of regular treatment of rivers, however, there was no obvious sign of serious upset in the river ecosystems, and certainly no sign of permanent elimination of any non-target groups. Nevertheless, some figures indicate a cumulative effect which could in the longer term present a problem.

In order to clarify this situation, the mass of sampling and surveillance data covering 5–6 years of regular treatment from 1974 to 1980 was analysed by two separate groups of freshwater biologists, namely the permanent team of French biologists (Dejoux *et al.*, 1980) and an independent team from Salford University, UK which analysed, on behalf of the WHO, the great mass of data collected in the project area and kept in storage in Geneva (Salford University, 1981). The latter team also reported on all the data available regarding the freshwater fish fauna in the project area, an essential part of the overall environmental study (Abban & Samman, 1980).

The reports from both assessment groups stressed the great difficulty in making definite conclusions that the reported faunal changes in populations of non-target invertebrates were clearly attributable to the larvicide treatment. This was due to the existence of many other natural factors influencing the numbers and distribution of fauna, and affecting the efficiency or otherwise of the different sampling methods used. Normal seasonal changes in the drift rate, natural variations in drift according to current velocities, uneven distribution of fauna in different parts of the same habitat, and considerable difference between different rivers in the extensive project area, all combined to produce a very complex picture which even the sophisticated factorial analysis employed by the French team could not clarify completely.

Nevertheless, the conclusions reached by the French team were that analysis of the bottom fauna samples, as judged by Surber samples taken from rock surfaces in particular and by artificial substrates, confirmed the existence of a long-term effect. This conclusion was based on data covering the most favourable period of the year for evaluation, namely during the months of falling river levels leading up to the end of the dry season.

On the basis of this data it was possible to define three arbitrary groups of non-target invertebrates according to their reaction to field treatments with Abate. Only one species emerged as highly sensitive, on a par with the target *Simulium damnosum*, this was the ephemeropteran *Tricorythus* sp. Among those only moderately affected were two hydropsychid species of *Cheumatopsyche* and the baetid *Pseudocloeon*, while among those least affected were several species of chironomids and a non-target species of *Simulium*, viz., *S. schoutedeni*. The report also concludes that in some genera of non-targets, difficulties in distinguishing closely related species in the immature stages might well obscure faunal changes at species level. This is well illustrated by finding that one species of non-target *Simulium*, namely *S. adersi*, disappeared from some treated areas to be replaced by another non-target, *Simulium schoutedeni*.

IMPACT OF FIRST TREATMENTS AND OF OVERDOSING

In rivers which have never been exposed to larvicide treatment, non-target fauna are liable to react sharply to the first application of Abate. As the continually expanding project area embraces new rivers or new areas of river, fresh populations of non-targets are continually being exposed to larvicide treatment for the first time. In order to mitigate the serious effect of this first impact the recommended practice was to carry out first-time treatments in new areas in the rainy season, when the increased volume of water facilitated rapid mixing of the chemical concentrate applied from the air. However, on one or two occasions, due to administrative or political pressure, the first treatment was carried out in the dry weather. The effect of one such treatment carried out on the river at the end of a long period of drought which had reduced the volume to the maximum degree, was further aggravated by gross overdosing with larvicide (Elouard & Troubat, 1979). The effect on the more-sensitive groups was to produce a catastrophic drift accompanied by complete elimination of standing crop (Surber and artificial substrate samples). Among the Ephemeroptera, the effect on populations of baetids, *Tricorythus* and leptophlebids was as lethal as on the target *Simulium* themselves, as demonstrated by their complete absence from drift collections 24 h after treatment.

In contrast, while the overdosing produced an almost equally high detachment of Trichoptera such as *Amphipsyche*, *Cheumatopsyche* and *Macronema*, this was not followed by any great reduction in the population attached to their rocky substrates; 24 h after treatment the drift pattern had returned to normal.

Over the whole project area it seems likely that such incidents involving gross overdosing could occur from time to time due to accident, miscalculation, or (as has been reported) unacceptably long intervals between the recording of river discharge and the actual larvicide dosing based on that figure.

INFLUENCE OF HABITAT AND 'ECOLOGICAL RESISTANCE' OF MACROINVERTEBRATES

Among the many factors contributing to the persistence of the non-target population after several years of exposure to Abate treatment, the precise microhabitat of different organisms must play an important part. Some species living in the same fast-flowing or turbulent rapids as the target *Simulium* larvae are more likely to be exposed to the maximum effects of the larvicide wave. Others which utilise the stream bed or the less turbulent backwaters tend to be 'ecologically resistant' because their habitats remain

199

relatively uncontaminated. This appears to be the explanation of observations made in an independent investigation within the project area on the effect of Abate treatment on the microcrustacean fauna (Samman & Thomas, 1978*b*). Laboratory tests showed that the predominant copepod *Thermocyclops hyalinus* was highly sensitive to Abate. But observations made on the drift samples in a partly impounded section of the White Volta River, at its point of entry into the Volta Lake in Ghana, showed that after a year of weekly treatments with Abate upstream from the sampling sites, the zooplankton still persisted in the reservoir.

It was considered that, apart from the possibility of drift from upstream populations, two other factors contributed to the continued survival of the microcrustaceans. One was the fact that this species also utilises the still water areas along the edge of the reservoir, thus escaping the mainstream flow of larvicide. The other factor is that their habitats also include the deeper parts of the reservoir (up to 14 m deep). From there the regular vertical migration to the surface, where they can be recovered by drift net, only takes place at night; the return to the deeper habitat by day enables them to escape the effects of day-time applications of Abate to the surface waters.

The independent report of the Salford University team who applied multivariate time-series analysis to most of the sampling data from the OCP project area, confirmed the French team's conclusions that the variables in Surber sampling were such that the technique should be restricted to the most favourable sampling period of the year. They were able to confirm that there were actually increases in the numbers of some non-target fauna such as Tanytarsiinae, *Simulium* larvae other than the target *S. damnosum* and Hydropsychidae, and also no clearly defined harmful effects on non-targets generally in the long run.

THE FUTURE OF THE OCP

It is important to realise that the aquatic monitoring programme will continue as long as the *Simulium* control project of the OCP exists (OCP, 1985*c*), not only with regard to rivers which have been regularly under treatment since the programme began, but also those from which treatment has been withdrawn and the many untreated rivers progressively included with the steady progress of the project area into western parts of the Ivory Coast. In this continually expanding project, new candidate *Simulium* larvicides which have emerged from earlier screening programmes, have qualified for large-scale field trials. While Abate is still the larvicide of choice in 80% of the project area, increasing attention is being given to its possible successors such as the organophosphorus compound chlorphoxim and the synthetic pyrethroids permethrin and deltamethrin (OCP, 1985*b*). The evaluation of these new chemicals will doubtless continue as long as the 20 year plus project lasts. Continuity of surveillance and uniformity of evaluation techniques is ensured by an annual meeting of international hydrobiologists convened by the WHO. Increasing attention is also being given to non-chemical methods of *Simulium* larval control; prominent among these is *Bacillus thuringiensis israeliensis* (Bti or Teknar) already being used on a large scale. Impact of such measures on non-target fauna will again be an important consideration, although this is outside the scope of the present review.

In the last 2 years, there has been increasing interest in supplementing the larvicidal control by measures directed against adult *Simulium* populations where they exist at high densities along the forested river banks, and also as a barrier against immigrating *Simulium* adults from uncontrolled areas outside the project. For this purpose synthetic pyrethroids are the insecticides of choice and had already been used for control of riverine tsetse flies in West Africa (see Chapter 8). In the adult *Simulium* control trials carried out in Togo, in the southeastern part of the OCP project area, the pyrethroids tested were permethrin and cypermethrin, applied directly along the river banks and associated gallery forest and other vegetation (Douben *et al.*, 1985). The effects of those treatments against non-target river fauna are discussed alongside the results observed when these compounds were applied in a similar manner against tsetse flies (page 166).

INFORMATION ON INVERTEBRATE REACTION
IN EXPERIMENTAL CHANNELS *IN SITU*

The miniature experimental channel, installed *in situ* in the natural river or stream (as described on page 56) has played a major role in the OCP in evaluating the impact of *Simulium* larvicides on non-target invertebrate fauna. The general purpose and procedure of this test has already been introduced; it remains to be seen exactly what type of information it has provided. As the main larvicide used in the first 10 years of the programme, and still used in 80% of the project area, was the organophosphorus compound temephos or Abate, the early series of trials were concerned mainly with this chemical, and with the comparison of its different available formulations. (Dejoux, 1975; Dejoux & Elouard, 1977; Dejoux, 1978; Elouard, 1984).

A series of experiments were carried out in these channels in two rivers which had a completely different history with regard to larvicide treatment. The first river, A, had never had any previous treatment, while the second, B, had been exposed to regular fortnightly treatment with Abate as part of the practical control routine. In River A 2 concentrations of Abate were tested, 0.025 and 0.05 ppm for 10 min while in River B the concentration tested was 0.1 ppm for 10 min. In each case an untreated channel provided the control. The drift collections were expressed quantitatively as drift indices, i.e. the number of organisms drifting per cubic metre per second. In the control channel, the indices ranged from 0.24–1.60 over 24 h, with a mean of 0.82. During that period the total number of drift organisms (270) represented only 3.25% of the total number of organisms exposed in the channel (viz., 8316).

In the channel treated at the rate of 0.025 ppm the drift over the following 24 h was 26 times greater than the control, with a peak drift index of 234. Results again showed that certain groups such as baetids and caenids (Ephemeroptera) and hydropsychids e.g. *Macronema*, agrionid dragonflies and chironomine larvae were particularly sensitive, all of them showing a peak of drift in the hour following treatment in what is now regarded as a classic pattern. Of even greater significance, certainly against the background of pesticide-induced drift in general, was the finding that the drift (although considerable) only represented part of the standing crop exposed to treatment. In this test the total organisms drifting over 24 h, 2894, represented only 27.4% of the total of over 10000 individuals exposed. After adjustment to allow for control drift in the same period, the figure is 24%. This must be regarded as a major contribution, or breakthrough, in the very difficult and contentious question about the precise relationship between drift and standing crop.

Tests at the higher concentration of 0.05 ppm, showed much the same pattern; the proportion of the 9872 exposed organisms in this case which drifted was 26%, only slightly different from previously.

In the third test carried out in a treated river, the Bandama, the preliminary control experiment showed a natural drift pattern very similar to that of the untreated river, *A*. But following treatment at the rate of 0.1 ppm for 10 min, a very different pattern was revealed. In the course of the following 24 h, drift amounted to only 11% of the total 4875 individuals exposed. Different reactions on the part of different groups still followed much the same pattern as before, with increased sensitivity on the part of caenids. The conclusion from these tests was that fauna already regularly exposed to treatment with Abate, showed a greatly reduced drift reaction to each new exposure, and a correspondingly diminished mortality.

This trend was confirmed in further tests in which fauna in the experimental channel *in situ* in the river were exposed to a very high concentration of Abate, namely 100 ppm for a very brief exposure of 5 s, a situation likely to occur in practical control operations at the point of larvicide application by helicopter. This overtreatment produced a very high detachment rate, viz., 47% of the total exposed by the end of 3 h, on the part of non-target fauna in the untreated river, whereas in the river regularly treated the figure was only 11.4%.

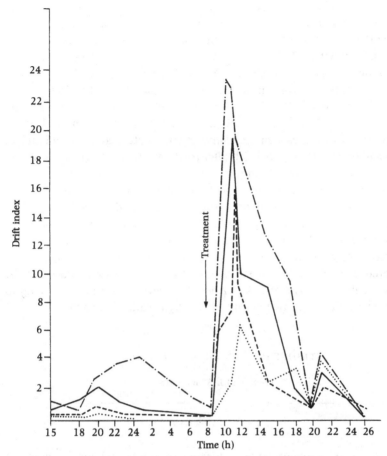

Fig. 9.5. Drift patterns of four main invertebrates groups following treatment of West African river with Abate at 0.1 ppm (after Dejoux & Elouard, 1977). The drift index is the mean number of drifting organisms per cubic metre of river water per second. Solid line, Chironomidae; dash-dotted line, Simuliidae; dashed line, Ephemeroptera; dotted line, Trichoptera.

The significance of these experimental findings is best illustrated by comparison with field studies being carried out in the same programme, using more conventional methods in order to assist in defining different drift patterns; for this purpose the ordinary drift nets, as used regularly to monitor the effects of Abate applications in rivers under regular treatment, were used (Elouard & Leveque, 1975). The drift nets in sets of 3 (see Fig. 6.2) were installed in the swift-flowing section of the river and operated for 30 min at a time at various intervals before and after treatment. A particularly instructive example was from a river in which regular treatment with Abate had been suspended for 2 months, allowing a heavy population of *Simulium* larvae to build up (Dejoux & Elouard, 1977). The effects of an application of Abate at 0.1 ppm on the four dominant groups of aquatic invertebrates (viz., *Simulium*, Trichoptera, Ephemeroptera and Chironomidae) are shown in Fig. 9.5. The figures show that within 30 min after treatment at 09.00 h there was a spectacular increase in drift, reaching a maximum drift index of 44 at 11.00 h compared with a figure of 1.02 immediately before treatment, and a maximum of 6.3 the previous night. The peak of drift occurred 1.5–2 h after treatment in the case of *Simulium*, Ephemeroptera and Chironomidae; with Trichoptera, the first peak 3 h after treatment was followed by a second lesser peak about 9 h after treatment.

Periodic drift sampling through the 24 h had previously demonstrated that all these organisms exhibit a nocturnal increase in drift as part of their natural behaviour. On the night following the treatment which had been carried out at 09.00 h the drift again showed an increase, but this time to an even greater extent than on the previous pre-treatment night. It is considered that this particular surge of drift represents those sections of the fauna in which the toxic effects of Abate had developed more slowly up to a point where the weakened organisms finally became detached.

STUDIES ON THE IMPACT OF ABATE

IN SMALL STREAMS

INTRODUCTION

Because of its great extent, its long duration, and unusual concentration of international scientists, it is natural that the great wealth of information emerging from the OCP should tend to overshadow the contributions made by lesser scientific investigations elsewhere. This impression could be further supported by the fact that so much of the work in West Africa has been concerned with Abate, and there might appear to be little more to be learned about the impact of that particular insecticide on the aquatic environment. However, the particular environment of the OCP study area, with its tropical climate and emphasis on large rivers, differs considerably from that in other regions of the world where environmental problems of Abate have also been investigated. These studies have been mainly concerned with a different range of *Simulium* species and associated non-targets from those in West Africa, and have in general been carried out in comparatively small *Simulium*-breeding streams rather than large rivers. In addition, the experimental procedures and the choice and application of sampling methods have followed quite independent courses from those in the OCP. The result has been to provide a novel approach to basic problems, complementing the experiences in the OCP, particularly with regard to work carried out in the USA and in Japan.

In the US studies in southern California, Abate was one of three blackfly larvicides tested simultaneously in the field, but for the moment only the Abate results will be discussed (Mohsen & Mulla, 1982). The site of this test was a comparatively small spring-fed creek about 4 km long to which Abate was applied at the rate of 0.1 ppm for 15 min to a riffle portion of the stream where *Simulium* larvae and non-targets were most abundant. Impact on the target *Simulium* larvae was measured by means of polyethylene strips which provided suitable artificial substrates for larvae and pupae. Non-targets were sampled by drift net (9 cm high × 30 cm wide) installed at mid-stream at a check station above the application point, and at 250 m downstream from the application point. Organisms in the drift were sampled for a 30 min period at 0, 30, 60, 90 and 120 min after injection of the larvicide. Surber samples of bottom fauna were taken at 5 sampling stations from 20 m above application point to 1300 m downstream from that point.

Four main groups made up the bulk of the drift organisms following treatment, and they revealed differences in drift pattern. The target *Simulium* larvae showed considerable drift by the end of the first 30 min, increasing sharply to a maximum of 19 000 per sample at 30–60 min, after which the numbers decreased but were still relatively high (774) 1.5–2 h after injection.

Among the non-targets, the numbers of chironomid larvae in the drift also increased with the first 30 min (40 per sample), but the maximum number (1210) was not recorded until the 90–120 min collection, i.e. later than with the peak for *Simulium* larvae. *Baetis* recorded the highest numbers in the drift, reaching a maximum of 1750 in the 90–120 min sample, but drift was slower to start, with none recorded in the first 30 min after treatment. Hydropsychids likewise showed a slower developing drift from 0 in the first 30 min to a maximum of 575 per sample in the 60–90 min period.

Despite the high drift rates, the Surber samples showed that the treatment had little adverse effect on the populations of chironomids and hydropsychids. In contrast, a marked effect was produced on the non-target baetids (as well as on the target *Simulium*), particularly evident 1 day after treatment. These effects appeared to be only transitory as none of the non-target fauna were eliminated by Abate treatment.

STUDIES IN JAPAN

One of the most penetrating studies on the effects of Abate on stream macroinvertebrates is that carried out by Japanese biologists in small tributary streams of rivers on Mount Tsukuba. There are many unusual features of these investigations designed to clarify the 'characteristic patterns in the processes of destruction and recovery of a stream ecosystem after an application of temephos'. The studies were not associated with any particular *Simulium* control programme in Japan but were clearly intended to provide more precise scientific data of value to *Simulium* control by larvicides in general, including the OCP. The small streams utilised provided the greatest possible contrast to the large rivers involved in the West African Volta Project, ranging in width from roughly 50–80 cm, with a depth of 15 cm, and discharge rates varying from about 2 l/s in one series to 6–12 l/s in another.

The first series of experiments were designed to define the minimum concentration of Abate which would induce drifting by *Simulium* larvae, and the reactions of non-targets to those concentrations (Yasuno *et al.*, 1981*b*). For this purpose, different concentrations were applied successively to the same section of stream. Treatment with 1 ppm Abate for 10 min was monitored by drift net sampling taken 10 m below the application point, at periods of 30, 60 and 90 min after application. At the end of that period, Abate was injected into the same section of the stream at the rate of 2 ppm for 10 min, and so on with increasing application rates of 5, 10 and 20 ppm for 10 min. The results showed that in general the total number of different species increased gradually with increase of concentration of Abate, but there were significant differences in the rate of reaction. This is well illustrated by two crustaceans, the amphipod *Anisogammarus* and the isopod *Asellus* (Fig. 9.6). Neither species showed drift reactions at 1 ppm/10 min. *Anisogammarus* started to drift at 2 ppm, and drifting reached a peak following treatment at 5 ppm, with up to 5000 individuals drifting in the first 30 min after treatment, the numbers falling sharply (to 100) in the following 30 min.

In contrast, *Asellus* did not respond to 2 ppm, but began to drift after treatment at 5 ppm, reaching a peak drift rate after the 10 ppm treatment.

Different drifting reactions were also observed among the four species of mayfly (Ephemeroptera) tested. Of these *Baetis* spp. proved to be the most sensitive, being one of the very few invertebrates drifting after the 1 ppm treatment, the peak of drift being reached at 5 ppm. Other species such as *Ephemera*, *Ephemerella* and *Paraleptophlebia* showed little response to 1 ppm and 2 ppm, and did not record maximum drift until the 10 ppm treatment. *Simulium* larvae themselves did not emerge as unusually sensitive; significant drift did not begin until 5 ppm, and this concentration also produced the maximum drift.

Effects on non-targets of control of blackfly larvae

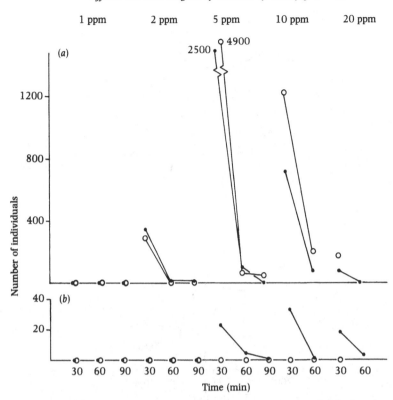

Fig. 9.6. Changes in the numbers of drifting individuals of (a) *Anisogammarus* and (b) *Asellus* after the application of various concentrations of temephos (Abate) in two tributaries of streams of Mt. Tsukuba, Japan, ○, ● (after Yasuno *et al.*, 1981*b*).

In a further series of experiments in this area of Japan a quite different procedure was adopted according to different objectives. (Yasuno *et al.*, 1982*c*). In this experiment drift nets were set up at 8 stations along the 1.4 km length of the stream in order to define drift patterns at different distances downstream from the application point. The tests were carried out in the winter months at a water temperature of 1.5 to 3.5 °C – in striking contrast to water temperatures of 25 °C or more regularly recorded in the West African rivers in the OCP. In view of the relationship between temperature and toxicity of Abate, the dosage rate in this series was 10 ppm for 30 min. Changes in the concentration of Abate downstream from the application point were ascertained by means of fluorescent dye which was also used to measure the time taken for the larvicide wave to reach the different downstream stations. The initial concentration of 10 ppm fell to 1.3 ppm 400 m downstream, and to less than 0.5 ppm at 1000 m, the

207

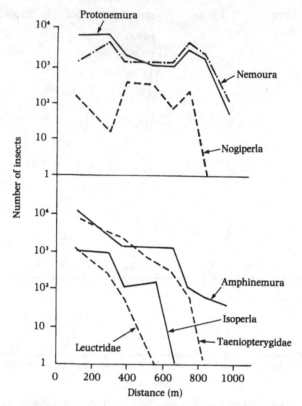

Fig. 9.7. Drift patterns of stonefly nymphs (Plecoptera) at different distances from application point following injection of temephos (Abate) at 10 ppm for 30 minutes to a stream in Mt Tsukuba, Japan (after Yasuno *et al.*, 1982c).

larvicide taking 5 h to reach that point because of the slow flow of the stream, 200 m/h.

The drift net data enabled three categories of drifting invertebrates to be defined.

1 More drifting during the first 5 h than in the following 15, e.g. *Baetis*, *Simulium*.
2 Numbers almost equal in the two periods, e.g. *Isoperla*, *Protonemura*.
3 Slow drifters, with drift numbers greater in the following 15 h than in the first 5, e.g. *Nemoura*, *Amphinemura*.

The drift net data also showed on the one hand the existence of different drift patterns according to distances downstream, and on the other that those patterns did not show any obvious relationship with the two categories above. This is well illustrated by the reactions of the different species of stonefly (Plecoptera). *Protonemura*, an intermediate drifter, and *Nemoura* a slow drifter, showed very similar drift patterns according to distance downstream, with drift numbers decreasing sharply between 800 and 1000 m (Fig. 9.7). With some species the drift numbers fell sharply to 0 between 600 and 700 m, e.g. *Perla, Chloroperla, Isoperla,* while with others such as *Amphinemura* there is a steady rate of decrease, with drift still taking place 1000 m below application point.

Similar differences in reaction appeared among different caddis flies (Trichoptera). *Hydropsyche* for example, which seemed to have only a low susceptibility to Abate treatment, showed diminished drift at downstream stations, whereas *Dolophilodes* spp drifted in large numbers at every station downstream to 1000 m.

The heavy treatment of larvicide in this series had a severe effect on most macroinvertebrates, especially in the region just below the application point; and, with the exception of *Hydropsyche* which recovered quickly, showed no signs of recovery until 2–3 months after treatment. Following the destruction of most of the macroinvertebrates there was a bloom of algae resulting from the disappearance of grazing fauna. This in turn led to sharp increases in the numbers of two groups of chironomids, (Orthocladiinae and Diamesiinae) within 2 months of treatment, which reached population levels higher than those prior to treatment.

This interrelated impact of Abate treatment on flora and fauna was the subject of a further investigation (Yasuno *et al.*, 1982*a*) following on the one above. In this series a different application rate was employed, viz., 2 ppm Abate for 3 h in order to clarify the impact on stream ecology as well as to compensate for the low water temperature of the streams. In this trial drift net studies were concentrated on the sampling station 240 m below the application point, drift collections being made at 30 min intervals over a period of 18 h. At the same time Surber samples of bottom fauna were taken in order to assess the effect of treatment on the standing crop.

The continuous drift sampling, in which more than 25000 individuals were captured, provided more information about the speed of reaction shown by different species to the wave of Abate. *Baetis* and *Isoperla* for example responded quickly with drift beginning 2.75 h after the application of Abate upstream. With *Paraleptophlebia* drift did not start until 7 h, and with *Nemoura* 8 h. However, the time of *peak* drift in these different groups was much the same, namely 9–10 h after application. *Simulium* larvae differed from these in showing a narrow margin between onset of drift at 3 h and peak drift at 4.5 h.

The simultaneous use of drift nets and Surber samplers revealed some apparent anomalies, illustrating the pitfalls of drawing conclusions from a single sampling technique. In the stonefly *Amphinemura* for example, which constituted about half of the 25000 individuals taken in drift nets, Surber samples taken prior to the experiment had yielded only 5 out of a total of 88 specimens captured. On the other hand *Protonemura*, which was the most abundant animal in the Surber samples, was scarce in the drift net collections.

In a further series of tests involving a comparison between Abate and fenitrothion the experimental procedure was again modified in an effort to compare these two chemicals under identical field conditions. This was done by applying the two insecticides to short sections of the same stream. A small stream was divided into 4 sections at intervals of 50 m. The experiment started with an application of Abate at 1 ppm for 10 min to the fourth section. One week later fenitrothion was applied at the second section. Drift samples taken at intervals of 10 to 20 min for 7 h after application provided further information about invertebrate reaction to Abate (Hasegawa *et al.*, 1982).

Among the most numerous drift organisms were *Baetis* and the amphipod *Anisogammarus*, the latter reaching a peak 20 minutes after application. In contrast, a caddis fly *Dolophilodes* did not begin to drift until $3\frac{1}{2}$ h after application, reaching a peak at 4–5 h.

Despite its rapid response and high rate of drift, *Anisogammarus* continued to be present in the treated section, providing further confirmation – already shown in laboratory tests – of its high tolerance to Abate (Yasuno *et al.* 1978). Those laboratory tests had indicated a mortality of less than 10% and 15% 24 h after exposure to 1 ppm for 10 and 30 min respectively.

A very similar experience which revealed a wide difference between the sharp behavioural response of an allied amphipod *Gammarus pulex* to Abate and its relatively high physiological tolerance to that chemical, has already been discussed earlier (page 64).

THE JOINT JAPANESE–GUATEMALAN
ONCHOCERCIASIS CONTROL PROJECT

In contrast to these intensive studies on impact of *Simulium* larvicides in small streams, it is worth noting that in one large-scale onchocerciasis control project involving small to very small streams, conditions are such that the reactions of non-target stream invertebrates to the larvicide treatment have only minor environmental significance. The programme in question is the Guatemala–Japan Cooperative Project on Onchocerciasis Control and Research which was established in 1975 and terminated in 1983 (Suzuki, 1983).

The main target species, *Simulium ochraceum*, breeds mainly in very small streamlets with a discharge rate in the range 0.25–0.29 l/s, and at current velocities which are much lower than those generally favoured by *Simulium* species in other parts of the world. As these streamlets are usually located in mountain areas, often heavily forested, aerial application of larvicide is impractical. Because of the reduced flow in these streamlets, ground application has to be carried out at numerous dosage points – only 50 to 100 m apart – at fortnightly intervals, and at much higher concentrations than those which have proved effective in large rivers. Under these conditions the carriage of the larvicide of choice, Abate, at a dosage rate of 2 ppm/10 min was found to be effective for only about 100 m downstream at a discharge of 0.8 l/s, and only for 25 m in a streamlet with a lower discharge of 0.13 l/s. Low carriage of larvicide is attributed to the absence of suspended matter from the clean clear water of these streamlets, and to the loss of effectiveness due to adsorption by stream-bed soils.

The Guatemalan streams infested with *S. ochraceum* are very small and poor in invertebrate fauna and fish life. Despite the high application rate of Abate therefore, it is considered that most of the chemical will be absorbed, with only a minimal amount flowing further downstream and reaching areas rich in invertebrate fauna and fish.

[1] The name *Simulium damnosum*, sensu lato, is used to cover what is now known to be a species complex, comprising several distinct species – of which there are at least six in the project area. A great deal of research is going on to clarify differences in behaviour and ecology between the different member species of the complex. In the context of the present review, the main interest of that work would be to find out to what extent larvae of the different member species are likely to react in different degrees to larvicide treatment.

TEN

IMPACT OF PISCICIDES

(AND MOLLUSCICIDES)

INTRODUCTION TO FISH TOXICANTS:

DEVELOPMENT OF SELECTIVE PISCICIDES

Fish toxicants are widely used to eradicate some or all of the fish in a body of water in order that desirable fish may be stocked, free from predation, from competition or from other interference from undesirable fish. Fish poisons have a long history of use in many countries but it is only in the last 40 years that the subject has really been scientifically investigated, and only within the last 20 that the full environmental or ecological effect of such toxic chemicals has been examined critically, particularly in the US and in Canada. Fish toxicants have been used in all types of water body, both static and running. Earlier progress in their study has been exhaustively reviewed (Lennon et al., 1971), and information available at that time regarding the reactions of freshwater fauna in general, including fish, to those fish toxicants was also the subject of a separate review at that time (Muirhead-Thomson, 1971). In the present review, space limitations would now make it extremely difficult to do justice to the mass of new information covering all freshwater fauna and all types of water body. Accordingly, in keeping with the scope of the coverage, progress since that time will deal only with the use of fish toxicants in running water, and only with the macroinvertebrate fauna at risk.

From the point of view of fish control, the running waters of streams and rivers pose an entirely different range of problems to those encountered in the static waters of lake and fish ponds in which so much of the earlier work in that field was concerned. These differences have already been stressed in the present review: in addition to the range of physical and biological characteristics of running waters, and the various factors which determine the transport of toxicant downstream, the factor of biological zonation is one of particular importance in fish control studies. In both Europe (Huet, 1959) and in the US (Lennon & Parker, 1959) dominant fish zones have been defined from source to mouth, and the implication of this at least is that a

212

problem pest fish against which measures have to be taken may inhabit only a section of the whole stream, making it necessary to define the upper and lower limits of its zone before effective control measures can be initiated.

One of the earliest motivations for destroying fish populations in streams was the control of disease, involving the elimination of all fish from springs providing the water for fish hatcheries. In this operation, two of the common piscicides were rotenone and toxaphene, which were also widely used in general control of rough fish, in stream reclamation and in its ultimate restocking with selected game and forage fish.

Considerable technical advances were made in the control of undesirable fish in streams with the successful development of a method for the selective poisoning of sea lamprey larvae in the tributaries in the upper Great Lakes where approximately 10% of the 3 000 tributaries are infested. The important selective chemical responsible for this breakthrough was TFM (trifluoromethyl nitrophenol) which had the unique property of killing larval lampreys at concentrations relatively harmless to game fish, wildlife and live stock. (Applegate *et al.*, 1961; Meyer & Schnick, 1983). Further research showed that TFM could be synergised by the addition of a widely used molluscicide, Bayluscide (Bayer 73) or niclosamide which had played an important worldwide role in the control of the intermediate host snails which harbour the parasitic worms responsible for human bilharziasis (Marking, 1972). Formulations of TFM with only 2% Bayluscide added (TFM-2B) were found to retain the selective action against larval lampreys, while having considerable advantages (Howell *et al.*, 1964) in other ways.

An almost equally important development in the 1960s was the introduction of the antibiotic fish toxicant, antimycin, which was found to have very wide application in the general field of stream reclamation, as well as exhibiting valuable selective properties in many cases between its action on game fish and that on trash and predatory fish (Lennon & Berger, 1970).

Critical field and laboratory studies on the reactions of macroinvertebrates to fish toxicants, and to streams treated with these chemicals, have been mainly concerned in recent years with these two main piscicides, TFM (and TFM/Bayluscide) and antimycin (Fintrol), but useful studies have also been carried out on the organochlorine compound Thiodan (endosulphan) – an insecticide already discussed in the context of tsetse fly control (see page 164). The extent of piscicide application in recent years is illustrated by the fact that between 1958 and 1983 approximately 1 265 000 kg of TFM and 8400 kg of Bayer 73 were applied to the tributaries of the Great Lakes to control sea lamprey larvae. This has led to some concern about the environmental effects of the repetitive use of lampricides, which have all been embodied recently in a comprehensive report on *TFM and Bayer 73: Lampricides in the Aquatic Environment* issued by the National Research Council of Canada (1985), which marks an important milestone in documentation on this subject.

THE LAMPREY LARVICIDE, TFM: EARLY

LABORATORY AND FIELD TESTS ON STREAM BIOTA

In one of the original classic reports on the discovery and application of selective lamprey larvicides (Applegate *et al.* 1961) it was reported that TFM was not harmful to the limited range of invertebrates included in stream tests. More precise information about the tolerance levels of macroinvertebrates at risk was provided by a series of laboratory tests in which representative species were exposed to different concentrations of TFM in static tests, mortality being recorded after exposure periods of 22–24 h (Smith, 1966, 1967). In the case of the leeches, snails and clams tested, these became sluggish when exposed to the chemical, making it difficult to define mortality at the end of the exposure period. Accordingly, at that point the test species were transferred to fresh water for a recovery period of 8–10 days. Simultaneous tests on sea lamprey larvae and on rainbow trout (*Salmo gairdneri*) were used as references to establish the concentrations adequate for lamprey control under these conditions.

Unfortunately, possibly due in part to trauma sustained in the process of collection of invertebrates from field sources, and to the trauma during handling and maintenance in the laboratory, control mortalities with many species were unacceptably high for firm conclusions to be drawn. Where these variables were minimal it was concluded that concentrations at which 100 % mortality was produced were 3 ppm for *Hydra*, 8 ppm for turbellarians and blackfly larvae, 12 ppm for burrowing mayflies, and 16 ppm for clams.

During the field application of TFM for sea lamprey control, concentrations are maintained between the minimum lethal concentration – MLC_{100} – which produces 100% mortality in lamprey larvae, and the maximum allowable concentration – MAC_{25} – which does not kill more than 25% of the rainbow trout (Howell & Marquette, 1962). In the present series of laboratory tests, the MLC_{100} was established as 4 ppm and the MAC_{25} as 10 ppm of TFM. It was concluded that stream treatment at the rate of 10 ppm would be expected to kill 100% of hydras, turbellarians and blackfly larvae. Burrowing mayfly larvae would be reduced by about 99% and clams by about 50%. Mortality among the other groups tested was reckoned to be less than the 25% allowable for fish. At the lower treatment concentration of 4 ppm, only hydras and turbellarians would be likely to suffer 100% kills, and only burrowing mayflies and blackflies would have kills greater than 25%. The general conclusion from this series of tests was that for most stream invertebrates, concentrations of TFM used for larvicide treatment did not constitute a hazard.

At about the same period of time an extensive series of field tests were carried out on macroinvertebrate reactions in treated streams (Torblaa, 1968). These were carried out in five stream tributaries of Lake Superior and four of Lake Michigan. One stream in each lake basin was used as a control, the other was treated with TFM at established larvicide dosages, while in another stream the treatment was 98% TFM + 2% Bayluscide. The average concentrations of TFM for each treatment were above the minimum lethal concentration for larval lampreys. Samples of invertebrate bottom fauna were taken in 30 cm² samplers, and included a very wide representative range of insect groups recorded from riffle areas and sand detritus areas one day before and one week after chemical treatment. In the case of other streams, including the one treated with TFM/Bayluscide, further samples were taken 6 weeks after treatment, and in some cases 1 year later. The general conclusion from the great mass of data recorded was that the total number of invertebrates was smaller 1 week after treatment than before treatment; increased somewhat 6 weeks after treatment, and had returned to pre-treatment levels 1 year after treatment. Invertebrates were more severely affected, and recovered more slowly, in areas of sand and detritus than in riffle areas.

INTENSIVE LABORATORY STUDIES ON THE IMPACT

AND UPTAKE OF TFM

A further incentive to obtaining more precise information about the reactions of stream invertebrates to TFM was provided by the fact that, under US legislation, petitions for the registration of such pesticides must include data describing its effect on the non-target biota, and on the fate of the chemical within treated animals and in the environment (Meyers & Schnick, 1976). Because of insufficient information of this kind at that time, the registration of TFM was cancelled in 1970 by the Environmental Protection Agency. Since that time extensions have been granted to allow time for the necessary investigations. As a result of this a range of stream invertebrates have been tested under rigidly controlled laboratory conditions, particular attention being given to the influence of water quality with regard to soft, hard and very hard waters, as it was already known that the toxicity of TFM to lamprey larvae themselves decreased as hardness and pH increased (Howell *et al.*, 1964).

The invertebrates selected for special attention included important fish food organisms, playing a key role in aquatic food chains, such as larvae of the midge *Chironomus tentans* and nymphs of the mayfly *Hexagenia*. They also included 'scud', *Gammarus pseudolimnaeus*, aquatic 'sow bugs', *Asellus brevicaudus*, damselfly nymphs, *Ischnura verticalis*, crayfish *Orconectes nais*, and the daphnid *Daphnia magna*.

215

The course of these toxicity tests illustrated very clearly the comparatively young and exploratory nature of this branch of environmental toxicology, in that the diversity of the invertebrate forms and their differing reactions to the artificial laboratory environment, makes it quite unrealistic to lay down standard procedures which are equally applicable to all. This phase of laboratory investigation is discussed in Chapter 3 in conjunction with experiences in other fields. It is sufficient to cite here the significant points emerging from the TFM tests. It was shown, for example, that in the case of the aquatic midge, *Chironomus tentans*, the TFM had an immobilising effect making it difficult to determine the exact concentration at which mortality occurred. The criterion adopted therefore was not the conventional LC_{50}, but the EC_{50} i.e. the concentration at which 50% of the test animals became immobilised. The extent of immobilisation at a range of concentrations was recorded at various periods of continuous exposure in a static test, viz., after 8, 24, 48, 72 and 96 h (Kawatski *et al.*, 1974, 1975; Kawatski & Zittel, 1977). However it was observed that some chironomids which were immobilised by TFM and appeared to be dead, revived when they were placed in toxicant-free water. By carrying out a series of such post-exposure observations it became clear that there was a wide discrepancy between EC_{50} and LC_{50} values. In soft water for example the 8-h LC_{50} was actually 12 times greater than the immobilisation-based EC_{50}. In longer exposures of 24–96 h, dead chironomids were more easily recognised, and the disparity between the two criteria diminished.

With regard to nymphs of the mayfly *Hexagenia* (Fremling, 1975; Fremling & Mauck, 1980) these organisms prefer silt bottoms where they can construct burrows. Accordingly, it was found necessary to design a special type of test vessel provided with an artificial substrate and each containing 10 burrows occupied by a single nymph, through which there was a constant recirculation of test water (Fig. 3.2). Straightforward records of mortality at each concentration enabled acute mortality to be expressed in the conventional LC_{50}.

With regard to other invertebrates tested (Sanders & Walsh, 1975; Sanders, 1977) acute toxicity tests were carried out by conventional 96-h static toxicity tests as well as by 30-day flow-through techniques; figures from the former test showed a 96-h LC_{50} of 57 mg/l for *Gammarus* and 110 mg/l for crayfish.

In addition to these, experiments were carried out on the uptake of TFM from water by 4 species of crustaceans and 2 immature insects at a range of concentrations from 0.013 to 0.510 mg/l using ^{14}C-labelled TFM, and showed that 6 species of invertebrates accumulate residues up to 58 times the concentration in water.

Earlier studies on fish had shown that with TFM the toxic effects were considerably less under conditions that simulated treatment of an actual

216

stream than the effects which were observed under static conditions. A similar trend was found to exist in the invertebrate tests, TFM being for example twice as toxic to crayfish in static toxicity tests than in flow-through tests. In a separate investigation on invertebrates (Chandler & Marking, 1975) TFM was also found to be more toxic to snails in static than in flow-through tests.

The tests with the burrowing mayfly *Hexagenia* showed that TFM was considerably more toxic in soft water than in hard water or similar pH and temperature, the 24 h LC_{50} being 4.75 mg/l and 6.50 mg/l respectively.

The general conclusion from all those exhaustive laboratory tests was that under similar test conditions TFM was toxic to invertebrates but even more toxic to larval sea lampreys. In trying to relate the great amount of data on the laboratory reactions of stream invertebrates to TFM, to actual field conditions, it is useful to remember that field application rates of TFM in lamprey control are usually at the rate of 2–4 mg/l, and that the normal period of application, 8–12 h, is much less than many of the standard exposure periods adopted in routine tests.

COMPARISONS OF TFM, BAYLUSCIDE

AND THE COMBINATION OF THESE (TFM/2B)

With the increasing use of Bayluscide (Bayer 73) in lamprey control either alone in some special situations or more commonly combined with TFM in the proportion of 98% and 2% Bayer 73, increasing attention was directed to the environmental aspects of these separate compounds or combinations. In practical control the combination of toxicants, TFM/2B exhibits greater selective toxicity towards sea lamprey, but is less toxic than Bayer 73 alone. The minimum lethal concentration required to kill 100% of the lamprey larvae, i.e., the MLC_{100}, is 1.0 mg/l, compared to a figure of 0.1 mg/l for Bayer 73 alone (Rye & King, 1976). The corresponding LC_{50} values were 0.78 mg/l for TFM/2B and 0.049 mg/l for Bayer 73. These latter values were the ones which could be directly compared with a range of LC_{50} values obtained in a laboratory study on acute toxicity of these two compounds to stream invertebrates.

These laboratory tests (Rye & King, 1976) were based on 24-h exposures, but in the case of snails and clams a further 96-h holding period in clean water was added to allow decisions to be made between anaesthesia and mortality. Although the tests were of the static type, compressed air ensured some degree of water movement and aeration.

In the case of Bayer 73, LC_{50} values in excess of 50 mg/l were recorded for crayfish, dragonflies, water boatmen, snipe flies and dobson flies (*Corydalus*). *Asellus* proved to be relatively tolerant, with LC_{50} of 23.0 mg/l, and also the burrowing mayfly *Hexagenia* with a figure of 6.9 mg/l. More susceptible were *Gammarus* (2.6), stoneflies (1.07), non-burrowing mayflies (*Stenonema*), 2.27, net-building caddis, *Hydropsyche*, 2.45, case-building caddis (*Helicopsyche*) 1.67, and *Simulium* larvae 0.255 mg/l.

The most susceptible organisms proved to be soft-skinned animals such as snails, *Physa* and *Pleurocera* with figures of 0.016 and 0.355 mg/l respectively, clam with 0.382, leeches with 0.42, *Tubifex* 0.034 and *Lumbriculus* 0.14 mg/l.

In contrast, all non-targets proved to be less susceptible to TFM/2B than to Bayer 73 itself, with the exception of *Hexagenia* (LC$_{50}$ 4.0 mg/l.) Turbellaria, *Physa* and *Tubifex* being again the most sensitive of all organisms tested. At this particular stage in progress, information was lacking about the effects of Bayer 73 alone on invertebrate communities in treated areas, but the indications from the laboratory tests were that the lampricide treatment at the rate of 0.1 mg/l would kill 100% of the turbellarians and *Tubifex*, 42% of *Physa* and 12% of the *Simulium* larvae.

IMPACT OF TFM AND TFM/2B IN HARD AND SOFT WATER STREAMS: INVERTEBRATE DRIFT AND LONG-TERM EFFECTS

Very early in the investigations on the effects of TFM on larval lampreys themselves it was established that water quality is an important factor determining toxicity, particularly with regard to hardness and pH, toxicity decreasing with increasing hardness. Up till this point the majority of field investigations had been carried out in soft water streams where practical TFM treatments are generally shorter in duration and lower in concentration than in hard waters. In recent years increasing attention has been paid to a more critical examination of TFM's environmental impact under those two contrasting water conditions. Equally significant have been the critical studies on the influence of TFM on invertebrate drift.

In Canada where these activities have been most marked, over 250 streams throughout the Great Lakes watershed are regularly treated with TFM. Each stream is treated once every three or four years, and exposure during treatment is less than 24 h. In soft waters the treatment is less than 3 mg/l while in hard water dosage has to be increased to 5–10 mg/l.

A study was carried out on the effect of TFM applied to two soft water streams, and the effect of TFM/2B applied to a hard water stream in one of the first critical studies involving both population studies and invertebrate drift (Dermott & Spence, 1984). In the soft water streams, previous tests had shown that minimum lethal concentrations of TFM at the rate of 1.1 and 0.8 mg/l for 12 h ensured complete mortality of the larval lampreys. Surber samples were taken in a riffle in midstream where depth, flow and sediment size were similar; and drift samples were collected daily for 3–4 days before and during treatment, and over the 4 days after treatment, the nets being set for approximately 2-h duration.

Three weeks after treatment, philopotomid caddis flies – *Dolophilodes* and *Chimarra* – were absent from all the treated riffles, whereas other dominant invertebrates had re-established themselves.

In the hard water community, treatment with TFM/2B produced a reduction in annelids, leeches being most affected. The day-time drift rates showed a catastrophic drift of leeches, and also an increase in drift of *Gammarus*.

It remained to be seen how the ecology of hard water streams was affected by treatment with TFM alone, and this was studied in a hard water tributary of Lake Ontario (Kolton *et al.*, 1985). Three sampling sites were selected in the experimental area, *a* Untreated, 1 km upstream from the application point; *b* and *c* Treated section, 0.9 and 4.8 km downstream respectively from the application point. The river was treated with TFM on 25 and 26 October as part of the routine sea lamprey control programme, maximum concentrations of 8.20 to 9.45 mg/l being maintained for 15 h at site *c*.

Benthos was sampled at each site prior to and following treatment, the post-treatment samples being collected 2 and 19 days, and 30 weeks, after treatment. Invertebrate drift was collected at site *c* for a 1-h period every 4 h, from 2 days before treatment to 2 days after, using three rectangular drift nets placed approximately equidistant across the stream. In keeping with the general experience that hard waters often contain more diverse fauna and abundant populations than soft waters, 68 taxa of macroinvertebrates were recorded in the three riffle sites.

An important factor in the evaluation of TFM impact was that the three different sampling sites showed significant differences in community structure before any treatment started. Chironomids for example accounted for a much greater proportion of the benthos in site *b* than in site *a* and the three sites showed differences in the proportions between such dominant forms as *Hydropsyche, Cheumatopsyche* and tubificids. In addition, there were considerable changes in the faunal abundance of the untreated site during and in the months subsequent to the experiment, and short-term changes associated with changes in river discharge. Studies on drift prior to treatment showed that its composition generally resembled that of the benthos at the nearby site – *c* – with *Hydropsyche* and tubificids being the most abundant members. Evaluation of drift had also to take into account that among the abundant Trichoptera some species were normally day drifters while others were night drifters (Waters, 1972).

The evaluation of TFM impact had to take into account all these variables, and this involved the analysis of a considerable amount of data. In general the TFM application did not affect the relative abundance of most of the river macroinvertebrate fauna, but increase in drift on the part of certain forms indicated that certain macroinvertebrates are sensitive to TFM and show an almost immediate drift reaction to treatment. Two of the most sensitive members were the trichopteran *Chimarra obscura* and the turbellarian *Dugesia* sp which showed large increases in drift abundance during treatment, and were subsequently nearly eliminated. These findings are of particular interest in that they show close accord with earlier laboratory findings (Smith, 1967; Maki *et al.*, 1975; Gilderhus & Johnson, 1980).

Most of the Rouge River macroinvertebrates adversely affected by the TFM lampricide treatment had begun recovering substantially 19 days after treatment, the rapid recolonisation being undoubtedly related to the hardness and pH of that particular water.

ANTIMYCIN, THE ANTIBIOTIC PISCICIDE:

IMPACT ON MACROINVERTEBRATES OF PONDS

AND RUNNING WATERS

The antibiotic piscicide antimycin A has now been a recognised fish poison for over 20 years (Lennon & Berger, 1970; Lennon *et al.*, 1971), and has been widely used in lakes and streams, particularly in the US and Canada. As fish differ in their sensitivity to antimycin, many opportunities are provided to practise selective control against target fish and over-abundant populations of 'trash' species in fish farms, ponds and other static bodies, (Burress & Luhning, 1969 *a,b*; Gilderhus, Berger & Lennon, 1969). The conclusion from all this earlier work was that at practical fish-control dosage, 3–10 ppb of antimycin, there was no significant effect on aquatic invertebrates (Marking & Chandler 1978).

Invertebrate reaction in ponds was later examined more critically at the Fish Pesticide Research Laboratory, Columbia, Missouri (Houf & Campbell, 1977) in experiments involving exposure to heavy applications of antimycin, viz., 40 µg/l. These ponds were rich in aquatic invertebrates with 74 animal taxa being recorded, dominant among these being mayflies (*Caenis* and *Callibaetis*), dragonflies (*Enallagma* and *Ischnura*) and a wide range of midges. None of these dominant groups were eliminated by the heavy antimycin treatment, nor were there any long-term effects on bottom fauna or interference with insect emergence.

Opportunities to study the impact of antimycin on running water fauna arose later in connection with a project in Wisconsin to eradicate non-game species of fish from a stream prior to restocking with brown trout (*Salmo trutta*) (Jacobi & Degan, 1977). The small stream selected for treatment ranged from 1.5 m wide at the upper section to 3.3 m wide at the lower end of the experimental section, with depth ranging from 12 to 18 cm. A very similar stream in the same tributary area provided an ideal untreated control, with a similar discharge rate in the range of 0.15 to 0.17 m³/s and a wide variety of representative stream macroinvertebrates. Benthos and benthic drift were sampled periodically for 5 months before treatment.

However, what should have been a crucial test on impact of practical field dosages of antimycin, yielded only inconclusive results because calculating errors and equipment failure resulted in gross overtreatment. Instead of the calculated rate of 10 µg/l for 3 h and 20 min, concentrations of 25 µg/l for the first 3 h and 40 µg/l for the next 6 h were recorded at the first sampling station just below application point and 50 m above an impoundment. At the second sampling station 800 m below the impoundment, a concentration of 17 µg/l for 2 h was recorded followed by a rate of 44 µg/l for the next 7 h.

Perhaps not surprisingly this gross overtreatment had a drastic effect on the stream invertebrates, with a considerable reduction in population within 2 days of treatment, as much as 100% in the case of *Baetis* and *Gammarus*.

Treatment produced a sharp increase in drift rates, reaching a maximum in the case of *Gammarus pseudolimnaeus* at the lower station, 12 h after treatment. High drift rates were also recorded for Ephemeroptera, Trichoptera and Chironomidae. Despite this, all invertebrates had returned to normal densities 1 year after treatment.

The reactions of the case-bearing caddis, *Brachycentrus americanus* are of unusual interest in that the larvae became disoriented after treatment, moving sluggishly over stones and vegetation, 50% eventually leaving their cases and dying.

As this overdosing was unplanned, the results have rather doubtful relevance as to the likely response of stream fauna to normal piscicidal dosages. The previous study on pond fauna indicated minimal impact, and the likelihood is that the results of the stream treatment can only be regarded as an exaggerated response to gross overdosing.

The experiment is salutory in that such errors, or combinations of errors, are liable to occur in any field experiment involving application of toxic chemical at a calculated rate. While such errors can now be more readily checked in most cases, there is ample evidence that in the absence of such checks serious errors can arise in the interpretation of field sampling data in terms of calculated figures alone.

THE ROLE OF BAYLUSCIDE AS A MOLLUSCICIDE

IN SNAIL CONTROL

Bayluscide (Bayer 73) has long played a dual role, being both an effective controller of larval lampreys and a practical controlling agent for the aquatic snails which are intermediate hosts of the parasitic worms causing human bilharziasis. At the time when I last reviewed the subject of the environmental impact of molluscicides (15 years ago) Bayluscide was only one of several promising molluscicides used in practical snail control. The intense interest in field testing these different chemicals at that time, under a wide variety of conditions in both static and flowing water, stimulated a great deal of scientific research on their environmental impact on freshwater ecosystems. For that reason, the amount of information available was sufficient to justify a separate chapter in that review. During the last few years, the situation has changed dramatically. While the control or reduction of bilharziasis by means of snail control has still an important part to play in practical disease control campaigns, it has proved a very costly one for many of the developing countries most affected. Interest in development and screening of new molluscicides has declined, and virtually the only molluscicide now available is Bayluscide (Bayer 73) usually referred to in molluscicide literature as niclosamide (McCullough *et al.*, 1980; McCullough & Mott, 1983)

Environmental studies on Bayluscide as a molluscicide reached a peak in the 1960s while later studies on alternative molluscicides such as Frescon (*N*-tritylmorpholine) (Corbet, Green & Betney, 1973), sodium pentachloro-phenate and organometallic compounds are no longer relevant to this review as these chemicals have gradually been phased out.

As Bayluscide remains virtually the sole molluscicide still in use – particularly in Egypt where several hundred metric tons are applied annually – its impact on non-target aquatic fauna is still a matter of environmental interest. However the bulk of new information about this aspect has been provided, not from the field of molluscicide usage, but from the comprehensive studies on this compound in its role as effective lamprey larvicide, either alone, or more usually, in combination with TFM. It is the work in that field which has now been responsible for the great advances in knowledge about its environmental impact, and consequently in this review Bayluscide as a molluscicide must inevitably be absorbed in the discussion of the more progressive field of piscicides.

222

ELEVEN

HERBICIDES AND AQUATIC INVERTEBRATES

INTRODUCTION

The use of herbicides to control undesirable plant growth first developed on a large scale shortly after World War II and has been extending rapidly ever since that time. With their increasing use in agriculture, forestry and water-way clearance particularly in developing countries these chemicals now rank alongside insecticides as major environmental contaminants (Balk & Koeman, 1984). The continuous monitoring programme of streams flowing into the Great Lakes over the last 10 years for example, has shown that herbicide use in agricultural land has now increased to such an extent that they now constitute more than half the total volume of pesticides used in agriculture (Frank et al., 1982). Even in the UK where there are unusually stringent regulations controlling pesticides in the environment – particularly with regard to natural water bodies – many of the long-established herbicides such as 2, 4-D, dalapon, dichlobenil and diquat have been cleared under the Pesticides Safety Precautions Scheme, 1973, for use as aquatic herbicides for control of submerged and emergent aquatic weeds, and for the control of vegetation along the banks of rivers and drainage channels (Ministry of Agriculture, Fisheries & Food, 1985).

Early recognition of possible effects on fish life of herbicides applied directly to water or contaminating water by run-off from agricultural land, led to very thorough laboratory investigations in the UK on fish toxicity, and established the relative lethal levels of about 20 common herbicides based on 24-h LC_{50} values (Alabaster, 1969). Since that time there has been an enormous increase in information about fish reactions to herbicides, all of which is outside the scope of this review. Corresponding studies on aquatic invertebrates have been much less systematic and comprise mainly isolated and restricted investigations in many countries. The bulk of this work on the impact of herbicides refers to the static water fauna of ponds, lakes and reservoirs, with only minimal attention to running water invertebrates.

As the total amount of information about invertebrates in static waters is itself rather limited, it is felt that the scope of this chapter can usefully be extended to herbicide impact on aquatic invertebrates in general. Although

223

much of such material is strictly outside the stated scope of this book, these static water studies could provide a useful basis for invertebrate reaction to herbicides and give a pointer to promising lines of future research on stream fauna.

Herbicides may reach fresh waters from many different sources, including run-off and leaching from agricultural lands. But the most immediate and drastic impact is provided by herbicides applied directly to water bodies for the destruction of undesirable aquatic plants, and the clearance of vegetation from waterways (Robson & Barrett, 1977: AAB, 1981; MAFF, 1985). From the very nature and use of aquatic herbicides it is clear from the start that possible effects on aquatic invertebrates have to be considered from two quite different aspects, namely the possible direct effect of the toxic chemical, and the indirect effects caused by destruction of vegetation which normally provides substrate and shelter for fauna. The destruction of vegetation itself could also act in another indirect way, namely by upsetting the ecology of the water body in such a way (either chemically or physically) as to interfere with the normal trophic relationships of different organisms.

The indirect effect of weed removal is well illustrated by early experience by the Tenessee Valley Authority in 1966 when the herbicide, 2, 4-D was applied over an area of 3237/ha for the control of water milfoil (Smith & Isom, 1967). Many representative aquatic insect larvae such as dragonflies (Anisoptera), *Leptocera*, *Caenis* and elmid beetles, which were all present pre-treatment, disappeared following treatment and remained absent for 12 months. It has been the general experience elsewhere that eradication of water milfoil eliminates a broad expanse of substrate suitable for colonisation by large populations of epiphytic insects.

REACTIONS OF POND AND LAKE INVERTEBRATES TO
DIQUAT, DICHLOBENIL, PARAQUAT AND ENDOTHAL

The indirect effects of aquatic weed destruction are also well illustrated by treatment of small ponds with dichlobenil, widely used in the US for the control of submerged aquatic weeds (Walsh, Miller & Heitmuller, 1971). The treatment produced maximum effect on pond weed, *Potamogeton*, which were all eliminated, and on *Chara* of which 80% were destroyed. This was followed by an algal bloom composed of various species such as *Anabaena*, *Spirogyra* and *Oscillatoria* possibly due to the upsurge of nutrients released by the dead plants. This in turn led to a sharp increase in zooplankton until about 1 month after treatment when levels of both zooplankton and phytoplankton declined suddenly with the reappearance of *Chara*.

In some cases, fauna closely associated with a particular target water plant may survive the destruction of their normal substrate by moving to other unaffected sites. This is well illustrated in ponds treated with diquat for control of *Elodea* (Hilsenhoff, 1966). Following the death and decay of *Elodea* within 7 days of treatment, fauna normally associated with it were recorded on shoreline vegetation composed of plant species unaffected by the herbicide.

Similarly, in an experimental treatment of a reservoir with paraquat, many of the macroinvertebrates associated with the particular plants affected by treatment, disappeared. But evidence that they had survived on alternative substrates was provided by the early colonisation – at reduced numbers – of reappearing *Chara* (Brooker & Edwards, 1974). Other experiments, with paraquat used for control of weeds in small lakes, also indicate that despite the massive destruction and decay of vegetation, there were no great changes in aquatic invertebrate populations that could not be attributed to seasonal fluctuations, and that 1 year after treatment there was no discernible change in species richness (Way *et al.*, 1971).

With regard to the immediate toxic effects of herbicides, this has been investigated in depth in relatively few cases, and almost entirely deals with static water invertebrate fauna. Four long-standing aquatic herbicides (diquat, dichlobenil, dalapon and endothall) have attracted particular attention.

Diquat was reported as early as 1960 as one of the most promising herbicides for use on aquatic weeds, and a great deal of information accrued about the effect of this chemical on fish. One particular study dealt with fish food organisms (Gilderhus, 1967) particularly with zooplankton. Acute toxicity tests were carried out under static conditions in order to establish LC_{50} values, at concentrations of 0.1 and 3 ppm diquat and extending over a period of 8 days These laboratory tests showed that an 8-day exposure to 3 ppm produced almost 100% mortality in *Daphnia*. Chronic tests were also carried out in concrete pools treated in three different ways, (*a*) a single treatment at 1 ppm, (*b*) 2 treatments of 1 ppm, 8 weeks apart and (*c*) 3 treatments of 1 ppm 8 weeks apart, the treatment being determined by the fact that diquat effectively controls weeds at 0.5 to 5.0 ppm. In the multi-treatment pools numbers of *Daphnia* were much lower than in the single treatment case, but bottom fauna appeared to be unaffected, as also were copepods.

In other pond treatments with diquat at the rate of 1 ppm to control Elodea (Hilsenhoff, 1966) no evidence of direct effect on invertebrates was observed, but there were obvious indirect effects due to destruction of habitat, the decline being particularly noted in the case of the amphipod *Hyalella*.

LABORATORY TOXICITY TESTS
WITH DIQUAT AND DICHLOBENIL.

Rather more precise information was provided by a comparative study of two herbicides, diquat and dichlobenil (Wilson & Bond, 1969) in which toxicity studies were carried out on five aquatic insects and one amphipod (*Hyalella*). Static tests were carried out to establish LC_{50} values at 24, 48 and 72 h. In the case of diquat, none of the dragonfly (*Libellula*) or damsel fly nymphs (*Enallagma*), or tendipedid midge larvae died after the 96-h exposure to 100 mg/l, i.e. they survived concentrations 40 times greater than the field application rates. Mayflies and caddis flies also proved relatively resistant to the chemical, but amphipods on the other hand were very sensitive; the 24-h, 48-h and 96-h LC_{50}s of *Hyalella* were of the low order of 0.58 mg/l, 0.12 mg/l and 0.048 mg/l respectively. In comparison, the LC_{50} values for all other organisms were greater than 100 mg/l.

In the case of dichlobenil, *Libellula* and *Enallagma* were again most resistant with a 96-h LC_{50} greater than 100 mg/l. Amphipods and tendipedids proved to be most sensitive, with 96-h LC_{50} values of 8.5 and 7.8 mg/l respectively. Dichlobenil was observed to have a narcotisising effect on all invertebrates, including dragonflies, which were immobilised at the end of 96-h exposure but subsequently recovered in clean water. Many of the caddis larvae which had survived the 96-h exposure had left their cases by that time. Many of these larvae survived when removed to clean water, but although they were provided with suitable material for case building, none succeeded in constructing a new case in the 2-week observation period.

The impact of diquat was also studied as a mixture with another herbicide, endothal, also widely used in aquatic weed control (Berry, Schreck & Van Horn, 1975). A reservoir treated with a mixture of the two chemicals produced concentrations of 0.11 ppm and 0.17 ppm of diquat and endothal respectively at a depth of 144 cm, at the shallow end of the reservoir. Concentrations at the deep end never reached toxic concentrations, even for *Hyalella*, but occasionally, areas of uneven chemical distribution produced 'hot spots' of higher concentration, up to 0.73 ppm, exceeding the lethal level of diquat for amphipods – viz., a 24-h LC_{50} of 0.58 ppm – in the shallower part of the lake.

COMPARATIVE STUDIES ON THE IMPACT

OF DIQUAT AND ENDOTHAL

In a further study on the environmental impact of diquat and endothal, two of the most abundant crustacean species in the lake scheduled for treatment were selected for special toxicity tests. These were the amphipod *Hyalella azteca* and the isopod *Asellus communis* (Williams, Mather & Carter, 1984). Preliminary evidence from lake treatment with herbicide had indicated a significant difference in effect between the two. The laboratory test equipment consisted of a water reservoir and an inflow–outflow aquarium allowing each test chamber of 0.6 l to be exposed to a slow but continuous flow of water, or herbicide dilution, at the rate of 12 l/day.

As the recommended dosage for both herbicides is 1–2 ppm, the test concentrations selected were 1, 3 and 10 ppm. The test disclosed clear differences between the effect of the two herbicides.

On the one hand endothal did not produce any significant mortality, which in the case of *Hyalella* was in conformity with previously established values for the 24-h LC_{50} as being greater than 100 ppm in the case of an allied amphipod, *Gammarus*.

Diquat proved to be much more toxic to these indicator organisms. At 10 ppm, the mortality of *Asellus* was 20%, and 100% in the case of *Hyalella*. On 120-h exposures, mortality of *Hyalella* was high at all concentrations from 1 ppm upwards. In comparison, the mortality of *Asellus* never exceeded 50% under these conditions.

IMPACT OF PARAQUAT IN SMALL LAKES

Another widely used herbicide on which comparatively few invertebrate-impact studies have been carried out is paraquat. In experimental treatments of a series of shallow lakes in the UK, no adverse effects on macroinvertebrates were noted, except possibly in the case of *Asellus*. It was not possible to state definitely whether this was attributable to direct toxic action or to deoxygenation following weed removal (Way *et al.*, 1971).

In connection with the treatment of a reservoir at rates of 1.0 and 0.6 mg/l, static tests were carried out for 8–14 days on a range of invertebrates, including the isopod *Asellus* and the mayfly *Cloeon*. Of these, *Asellus* emerged as the most susceptible, but it was concluded that although it would seem that paraquat exerts a direct toxic effect on this organism at concentrations normally applied in the field, such concentrations would have to be maintained for several days to exert a lethal effect, while in fact the concentration of paraquat in heavily weeded ponds falls off rapidly (Brooker & Edwards, 1974).

EFFECTS OF CYANATRYN ON DRAINAGE CHANNELS

One of the significant trends in herbicide use in recent years has been the search for more persistent chemicals which would allow longer intervals between treatments, and which would also be better suited for application to running waters. The environmental impact of one of these new compounds, cyanatryn, has been studied in drainage channels in England (Scorgie, 1980). After treatment at the rate of 0.2 µg/g, a sampling programme was designed to deal with the invertebrate communities associated with each dominant plant (*Myriophyllum, Vaucheria* etc.). Once the plants in the sections of channel affected by cyanatryn had disintegrated, the resulting detritus was sampled instead, until recolonisation occurred. The effect of treatment was the replacement of a community dominated by a single species, *Limnaea peregra*, by a detritus-based community in which several groups or species were more or less co-dominant. Initially *Lymnaea* accounted for 81–84% of all species, but 4 months after treatment this proportion had fallen to below 10%, by which time there had been a marked increase in such organisms as *Asellus, Chloeon, Caenis*, chironomids and oligochaetes. Some groups, such as Corixidae, disappeared completely following cyanatryn treatment, but reappeared with the return of aquatic macrophytes one year after treatment.

PHENOXY HERBICIDES IN FOREST STREAMS

Despite considerable research on new chemicals, a dominant role is still being played in weed control after 25 years by the phenoxy herbicides, 2,4-D and 2,4,5-T (Brooker, 1976; AAB, 1981). In the US in particular, these are still widely used in forestry, agriculture and natural waters, and 2,4-D has been registered for use in irrigation systems (Boyle, 1980). Since the late 1960s, when 2,4-D and 2,4,5-T were used in Vietnam there has been increasing concern and controversy about their role as environmental contaminants, particularly after the finding in 1969 of the highly toxic contaminant TCDD in 2,4,5-T.

The environmental presence of these two compounds has received particular attention in the forest areas of Oregon, US (Norris, 1981; Norris, Lorz & Gregory, 1983) and these studies have yielded a great deal of information

228

about the fate of these chemicals in forest streams. Although the work was not directly concerned with macroinvertebrate fauna, it was able to define more clearly the range of herbicide contamination to which stream fauna might be exposed. Herbicides enter forest streams in several different ways; by direct application to the stream surface: by accidental drift from treatment units: by mobilisation of herbicide deposited in dry ephemeral field channels: by overland flow during high precipitation, and by leaching.

Direct application produces the highest concentration, but only for a short time. Concentrations only rarely exceeded 0.1 ppm, but higher figures up to 1 ppm are liable to occur in streams flowing from treated marshy areas. In such cases 2,4-D may be detected in the water for several days. Otherwise, residues disappear quickly in 1–2 days. This contrasts with the much longer persistence of this herbicide, up to 7 days or more, which usually occurs in the static water of rice fields, lakes etc.

Under favourable climatic conditions, such as the lower temperatures experienced in the Canadian winter, 2,4-D may persist for much longer periods in the water. In small, shallow artificial weed-infested ponds treated with a granular formulation (Aquakleen) of 2,4-D, residues in the pond water rapidly declined, but were still present at concentrations of 1.0 mg/l after 85 days, and at 0.2 mg/l after 6 months (178 days) (Birmingham *et al.*, 1983; Birmingham & Colman, 1985).

TWELVE

SUMMARY AND ASSESSMENT

From the very wide range of activities dealt with in this review, dealing with a great variety of problems in many countries, it is clear that the period under review has been one marked by a great upsurge of interest, and by remarkable progress in a hitherto much-neglected aspect of running water contamination. In fact this period of intense interest in the special problems of the macroinvertebrate fauna (as distinct from those of freshwater fish) of streams and rivers, may indeed encompass a peak period of study which is perhaps already on the decline. This is perhaps inevitable. Problems of particular interest or importance, or ones for which unusually ample funds are available for research, attract and encourage the most competent researchers, whose interest and scientific dedication in turn engenders further enthusiasm (Brown, 1973). This leads to the build up of multidisciplinary teams capable of making massive contributions to knowledge during their tenure.

This has certainly been the case with some of the major projects examined in detail in this review. For example, the OCP (Onchocerciasis Control Programme) inaugurated in 1974 on a 20-year basis has produced team work of the highest productivity during its first 10 years. Key research workers tend to move to other fields, or to retire; the teams become disbanded and funds for research tend to dwindle. It is difficult to visualise that the same intense research effort on the impact of *Simulium* control measures on non-target macroinvertebrates, and on stream ecosystems, will continue at the same high pitch for the next 10 years of the programme.

These forebodings are given further substance by the fact that one of the important long-term objectives of the OCP is for the programme to undergo a process of devolution in which the participating states of West Africa will progressively undertake certain responsibilities within their national structures for multidisease control. Any falling off in standards of scientific investigation and surveillance as a consequence of devolution could have unfortunate, if not disastrous results. There is already some evidence that the Abate treatments of rivers in the OCP area have had a cumulative effect over the years. If this is confirmed, and the ecological balance has been irrevocably upset, migrating *Simulium* adults which may in the future be able to penetrate areas from which spraying has been withdrawn, may find their

230

larval habitat depleted of its normal controlling predators, creating ideal conditions for the explosive proliferation of the target blackfly.

Perhaps the same trends are evident in the great peaks of intense environmental research in recent years on the impact of spruce budworm control operations on forest stream fauna. For the moment, knowledge about the impact of such widely used insecticides as fenitrothion (Accothion), carbaryl (Sevin) and permethrin has reached a stage almost of saturation, where an inevitable lull and period of stocktaking has already set in.

The same phase of stocktaking following a period of several years intense research on pesticide impact appears to apply to the Athabasca River project, one of the most productive scientific investigations in recent years. Critical reappraisal of study methods used, and in the interpretation of sampling data regarding macroinvertebrate fauna, has emphasised the need for a new and more critical approach to the problem of long-term repercussions of continued application of chemical larvicides to large rivers for *Simulium* control (W. O. Haufe, personal communication).

In view of all these trends it might be an appropriate function of the present review to indulge in a similar type of stocktaking, especially with regard to the whole methodology of evaluation which is the continuous theme from chapter to chapter. The need for such an overall assessment of progress also takes on some degree of urgency in view of the undeniable evidence that the world-wide use of pesticides, far from diminishing, is actually on the increase (Balk & Koeman, 1984).

All the information provided by the reports from the many different research workers quoted points to the essential need in all evaluation programmes to have a closely integrated field and laboratory phase of investigation. This integration can be most successfully achieved if the laboratory phase is planned with a full awareness of all the advantages, as well as the disadvantages, of applying laboratory experimental findings to the course of events in natural habitats. At the same time the limitations of conventional field sampling techniques – Surber samplers, artificial substrates, drift nets etc. – need to be more fully recognised and checked experimentally where possible under the wide variety of conditions in which they are to be used.

Without this constant awareness of sampling error and bias, the sampling data on which so much reliance is placed may in some cases be giving very misleading ideas about the impact of pesticide chemicals on stream macroinvertebrate fauna.

Five major pest control projects which have contributed so much valuable material for discussion in this present review, and which have all been equally concerned with the problem of assessing the effects of particular toxic chemicals on stream ecosystems, have each exhibited a quite independent approach to the relative roles of field and laboratory research. In the classic studies on environmental effects of insecticides used practically or experimentally for control of spruce budworm in eastern Canada, the very high standards established in field studies, first with fenitrothion and later with permethrin, contrast with the almost complete absence of back-up laboratory data dealing with invertebrate response to the range of pesticide contamination experienced in these projects.

A similar imbalance characterises the field studies on the impact of insecticides used in tsetse control programmes in West Africa and in Botswana, although, in all fairness, it must be pointed out that in the rugged terrain of many of these tsetse control areas, laboratory facilities for carrying out precise studies on macroinvertebrate reaction do not exist.

The West African Onchocerciasis Control Programme (OCP) into which such an enormous scientific effort has been invested, amply financed from international sources, is again one in which a precise laboratory phase of evaluation has lagged far behind the great progress made in the field. In this particular case, the gap has been filled to a large extent – sufficient for immediate practical needs at least – by the use of an experimental field technique, the experimental channel *in situ*, which provides some of the exactitude of the laboratory in the way of replicated exposure of test organisms to controlled time and dosage of the particular pesticide.

Another major *Simulium*-control programme, namely the Athabasca River Project in Canada has also been characterised by advanced field studies not supported by appropriate macroinvertebrate investigations in the laboratory.

In contrast to all of these are the environmental studies on the lamprey larvicides, TFM and Bayluscide, in the Great Lakes. In this programme very high standards of laboratory testing were established at an early stage, dealing with a wide range of stream macroinvertebrates. The laboratory techniques embody very precise standards with regard to uniformity of test conditions – such as temperature, pH, hardness, and other water quality factors – and were also integrated with an expanding field evaluation programme.

Disparity of approach in different projects is also evident in the extent to which they have utilised advances in physicochemical techniques for measurement of pesticide residues. These techniques have been fully exploited in most of the American and Canadian programmes reviewed, providing a massive background about the exact concentrations of pesticide in the affected water bodies in relation to the spraying regimen. This has enabled vital information to be obtained about the concentration profiles downstream from application points, and about the persistence of pesticides and their degradation products in the various components of the environment, including some stream macroinvertebrates themselves.

This physicochemical knowledge has also been well utilised in evaluating the impact of insecticides used for control of riverine tsetse flies. In Botswana for example it has provided valuable data of wide significance in other fields about the precise relationship between aerial application dosage rates and the exact concentration pattern in the affected water. Equally valuable information of wide significance has been provided in that project regarding downwind and drift dispersal of pesticide from the target area.

In striking contrast to all this is the almost complete absence of such chemical monitoring data from the Onchocerciasis Control Programme (D. Kurtak, personal communication, 1985). The scientists involved in that project are fully conversant with the repercussions of overdosing and the possibilities of wide gaps between the real and the calculated amount of larvicide chemical reaching the surface of the rivers exposed to aerial spraying. Nevertheless, chemical monitoring of Abate (the main larvicide used in the first 10 years of the programme) has been almost completely absent from the scientific investigations apart from an early isolated study on residues in mud and fish (Quelennec *et al.*, 1977). Consequently the field-monitoring data in this project, as distinct from the controlled experimental studies in the channels *in situ*, have all had to be related to a calculated concentration of pesticide in the river water in the absence of any direct measurements. There is ample evidence to suggest that in the absence of direct chemical monitoring, ideas about concentration profiles are highly speculative and (on occasion) grossly misleading.

Much of the discussion in this review centres round the theme of evaluation, with emphasis not so much on facts reported in the literature, but on the methods whereby basic figures have been obtained in the first instance. The validity of test methods and sampling techniques is of paramount importance in the interpretation of all reported 'facts' bearing on the environmental impact of pesticides on running water macroinvertebrates. Once again it is important in this discussion to consider laboratory evaluation and field evaluation as separate entities, while recognising that there are certain borderline semi-field areas where laboratory-type experiments have been carried out within the natural running water habitat itself.

With regard to laboratory investigations aimed at establishing tolerance levels of macroinvertebrates to different chemicals and to determine acute toxicity data, there has been marked change and progress in the last 15 years. The most striking feature has been the break away from the earlier rigid standards largely imposed by the quite distinct discipline of fish toxicology (Sprague, 1969, 1971). Nearly all the earlier studies on acute toxicities of aquatic macroinvertebrates adhered closely to the 48 h, 72 h, continuous exposure to toxicant in order to determine LC_{50}s. Now it is appreciated that methods which are appropriate to freshwater fish do not necessarily have any relevance to the totally different physiological reactions of aquatic invertebrates, or to the very wide range of organisms included in that category embracing a wide range of habitat requirements. In one direction the laboratory evaluation of macroinvertebrates has recognised that rigid standardisation does not allow for the fact that different organisms have different requirements which must be incorporated in the tests in order to maintain control mortalities at an acceptably low level. In order to allow for these differences in requirement, laboratory testing has branched out in many different directions in recent years. There has been less preoccupation with the idea that determination of LC_{50} values is the ultimate goal of these tests, and rather more attention to determining the reactions of organisms to the range of concentrations and time exposures likely to be encountered in the actual contamination of their natural habitat in the field.

In addition, many laboratory experiments are now designed in such a way that after the organism has been exposed to different pesticide concentrations for different time periods, it is removed to clean pesticide-free water for continued studies on its reactions. The prolongation of this type of test has disclosed a whole new concept about delayed mortality, and a new critical approach to what constitutes lethal levels and lethal concentrations.

In another direction, there has been increasing recognition that aquatic macroinvertebrates which live in running water environments must be laboratory-tested under flowing-water conditions if these tests are to give a realistic interpretation of their reactions under pressure of pesticide. With this in mind, a wide range of flowing-water tests have been devised by different workers, ranging from relatively slow flow or interchange of water at one extreme, through intermediate tests involving rapid through-flow, and culminating in laboratory-simulated streams or experimental channels which permit the determination of tolerance levels under conditions akin to those in the natural habitat. They also enable such behavioural reactions as irritability, detachment and downstream drift to be determined under controlled and reproducible conditions. In a further extension of this realistic approach, there has been the development of longer experimental channels, often outdoors, in which model running water communities can be established, permitting community response to pesticide impact to be observed under controllable conditions.

It seems significant that many of these new developments in techniques have taken place in laboratories not necessarily associated with pest control programmes, but rather in the nature of pure scientific research. A closer integration between these advances in laboratory research and the conditions which actually exist in natural habitats affected by pest control programmes, is clearly desirable. With such closer integration, the laboratory worker could adapt or devise techniques in accordance with the range of conditions to which stream fauna are exposed to pesticide contamination. Equally beneficial would be for the field biologists to define the type of information they require in order to facilitate the interpretation of field sampling data aimed at measuring invertebrate response. A flexible and adaptable laboratory programme, free from the restrictions and demands of whole-time commitment to 'standard laboratory tests' would be in an advantageous position to examine each problem experimentally in a manner best suited to the evaluation needs of the programme as a whole.

The essential need for integration between laboratory and field phases of evaluation has been expressed in essentially the same way by other authorities. Considerable interest has been aroused for example by the 'sequential hazards evaluation scheme' (Duthie, 1977; Tooby, 1981), which visualises a series of steps, each one of which generates more complex information. Five major steps are proposed: (a) preliminary screening; (b) more detailed acute toxicity testing; (c) chronic toxicity testing; (d) bioaccumulation and (e) environmental evaluation. The preliminary screening phase is usually carried out by the chemical manufacturers themselves, and the extent to which the follow-up steps are carried out will depend on the potential risk to aquatic fauna. The scheme is flexible in that it does not stipulate a strict sequence of test phases. At any stage new information coming to light may require earlier phase tests to be repeated, extended or modified in order to clarify particular situations. At each step in a hazard-assessment sequence, the options available and the decisions taken will depend on the nature of the information provided by earlier steps.

As mentioned above, early laboratory studies on pesticide impact on aquatic macroinvertebrates were heavily influenced by the methodology of fish toxicity studies current at that time. The rather slavish adoption of 'standard toxicity tests' and the undue preoccupation with the median lethal concentration (or LC_{50}) as the ultimate goal of laboratory experiments has long had a stultifying effect on the development of experimental methods to measure invertebrate response. That phase happily now appears to be in the past, as is clearly evident from the advances dealt with in this review. Significantly, it also appears that fish toxicologists themselves have had some heart searching in recent years, and many of the long-standing concepts have been very critically re-examined (Brown, 1976; Solbe, 1979, 1982; Alabaster & Lloyd, 1980), so much so that future developments in macroinvertebrate studies could greatly benefit from a full awareness of these redefined aims and objectives in fish toxicology research.

In the present review a great deal of attention has been drawn to the relevance of laboratory findings to field conditions, and to the design of naturally colonised artificial streams designed to bridge the gap between laboratory and field. This is also a problem which has received a great deal of attention in fish toxicology, where work on these different lines is contributing towards the determination of water quality criteria for freshwater fish (Solbe, 1979, 1982). Although much of that work is concerned with those contaminants to which fish are particularly susceptible – cyanides, ammonia, phenols, and metals such as zinc, copper and cadmium – the experimental approach has great relevance to macroinvertebrate biologists facing allied problems in evaluation.

The UK Water Research Centre for example, which has been particularly active in this field, has carried out artificial-stream experiments which are regarded as 'a compromise between the controlled but unnatural laboratory study and the natural but uncontrolled real situation of the polluted environment' (Solbe, 1979). The principal function of this study in an artificial stream was to establish the survival, growth rate and metal accumulation of the fish; but detailed study of invertebrates was also included, partly to try and define the pathway by which metals reach fish tissues in a semi-natural environment. The experience gained in 2 subsequent 3-year periods regarding the management of such a large experimental facility, and the use of sophisticated equipment for measuring the physiological reactions of fish to a wide variety of potential pollutants in the form of a continuous automatic alarm system, could all provide valuable guidance in future trends in macroinvertebrate studies. Finally, in the continuous progress towards the establishment of improved water quality criteria, it is clear that biological monitors are now being considered to be more reliable and realistic than the widely used physicochemical monitoring systems (Poels, 1975; Solbe, 1979). So far the emphasis has been on the rainbow trout as a suitable biological indicator, which exhibits a range of distinct physiological and behavioural responses to contaminants, responses which can be detected and amplified by electronic equipment. It would be encouraging to think that in those situations where the contaminating agents are pesticides, the untapped potential of highly sensitive macroinvertebrate species as biological monitors will receive due attention.

On this general problem of the role of simulated streams, it appears that until many more critical studies have been carried out, this will continue to be rather a controversial issue which has recently been comprehensively reviewed from all aspects (Shriner & Gregory, 1984). It is not possible to do full justice to that authoritative report within the confines of this book, but perhaps some of the most signficant statements can best be quoted *in extenso*:

The stream ecosystem includes the entire stream length from headwater to mouth. Because of this complexity, streams are considered the most difficult freshwater systems to model.

Although artificial streams seem to provide the link between laboratory and field situations, great caution must be used in the interpretation of results.

As the artificial system becomes more like the natural system, the ability to establish cause and effect relationships and mechanisms decreases since control and reproducibility decrease...The once-through water flow design permits the input of toxicants as well as nutrients to parallel that of the natural system...In order for an artificial stream to adequately simulate a natural system it must be multitrophic providing various functional group interactions as found under natural conditions.

It is clear that in designing an artificial stream system for use in toxicological research, compromises will be made between complexity and more simple systems with high degrees of reproducibility and control...There is currently no standard model stream nor a standard procedure for testing toxicants...While the outdoor system, especially stream-side or in-stream flumes, more closely resemble the ecosystem to be modelled, it is more difficult to recognise toxicant effects separate from ecosystem variability...One reason for using communities is to observe the interaction of toxicant effects with community dynamics. This is particularly useful if laboratory studies have previously been done on single species...Even though laboratory streams can be utilised to investigate a large array of environmental perturbations, their reliability to predict effects in nature is questionable.

When we come to assess progress made in evaluating pesticide impact in the *field*, it is evident that progress in the last 15 years has been much less dramatic than in the sphere of laboratory experimental work. A relatively small number of conventional sampling methods for stream macroinvert-ebrates forms the essence of measuring the effect of pesticide pressure on the macroinvertebrate community. Dominant among these methods are the Surber sampler for bottom fauna and drift nets for measuring pesticide-induced benthic drift. These techniques are supplemented on occasions by a variety of artificial substrates and, in some instances, by the use of caged indicator organisms. Despite the great effort which has been put into these field projects, in which routine sampling of stream invertebrates often in-volves round-the-clock collecting as well as the physical hazards of working in fast-flowing rivers, there has been surprisingly little critical examination of the efficacy of these basic sampling methods under different operational conditions.

In the field of pure freshwater biology, sampling methods used for macroinvertebrates are constantly under review with the aim of determining the advantages and limitations of widely used techniques (Brooker & Morris, 1980; Furse *et al.*, 1981; Drake & Elliott, 1983; Mackey, Cooling & Berrie, 1984; Wright *et al.*, 1985). In the field of pesticide evaluation it appears that in most cases the design of sampling procedure is selected more or less arbitrarily, early in the project, and continues to be used thereafter. An outstanding exception to this widespread approach is provided by the experimental work carried out by Canadian biologists on the efficiency or otherwise of conventional drift net designs; critical experiments which led to the 'bomb' design specially suitable for rapid-flowing water, and eventually to a further improved design suitable for large fast rivers such as the Athabasca.

The superior design of these modifications was firmly established by comparative tests with conventional conical drift nets under identical conditions. This is a practice which could be emulated with advantage in many other projects involving running waters. Some preliminary experimental work as to the design of drift net most effective under local conditions of water flow, stream depth etc. could well lead to modifications which could counteract or minimise inherent 'trap bias'. These designs might also enable samples of live benthos to be obtained in such a way that they remained free from damage by the trapping technique itself, and could consequently be retained for further study on delayed or long-term effect from the exposure to pesticide.

The amount of research effort which goes into the problem of evaluating pesticide impact in the field, and the extent to which continued research enables a sound basis of information to be built up, are greatly influenced by a factor over which the investigating biologist may have little control. This is the length of time over which a particular pesticide is considered the most suitable for a specific pest control problem before it is replaced – for reasons of improved effectiveness, lower cost or other reasons – by another pesticide which may present entirely different environmental problems. Some of the major pest control programmes have been dominated by a single chemical pesticide over periods of several years. In the OCP in West Africa for example, the first 10 years of the programme, up to 1985, were dominated by temephos (Abate). Similarly, *Simulium* control in Canada over roughly the same period has been almost entirely dominated by methoxychlor. In the control of sea lamprey larvae also, in the Great Lakes area, TFM alone or in combination with Bayluscide continues to be the pesticide of choice after nearly 25 years.

In contrast to these examples of stable continuity, the situation in Maine, US (the scene of so much intensive environmental study) has undergone several changes in the choice of insecticides used for spruce budworm control (Nash *et al.*, 1971; Stratton, 1982–84). In the post-DDT era, fenitrothion (Accothion) played an important role up till 1975, since when it has only been used experimentally. Since that time carbaryl (Sevin) has been the main insecticide of choice, supported by acephate (Orthene). Since 1982 the state has increased its use of the insecticide Matacil (aminocarb) as well as Zectran, and there has also been an increase in use of the biological control agent, *Bacillus thuringiensis israeliensis* (Bti). In 1985 priorities were affected by an unpredictable event in another continent, namely the Union Carbide incident in Bhopal, India; as an indirect result of this, Zectran became unavailable, and 80% of the project came under Bti control, which was chosen over Matacil, fenitrothion and Sevin (J. G. Trial, personal communication, 1985).

Changing situations like this can quickly render the best efforts of environmental biologists outdated or of no further relevance; nor do they provide any incentive to plan long-term research projects based on a specific pesticide. However, although it might appear that sustained research on a particular chemical pesticide must inevitably cease to be of value when that particular chemical is withdrawn from practical field use, the experience gained in evaluation makes an indelible contribution to the sum total of knowledge about the measurement and assessment of pesticide impact in general. Furthermore, the build up of experience with pesticides also provides a methodology which can be applied to other 'perturbations' of the stream environment, whether chemical or otherwise. In the last few years for example, there has been a great increase in biological agents in pest control affecting running waters. Bti (*Bacillus thuringiensis israeliensis*) is being used increasingly in the OCP as an alternative to organophosphorus larvicides, and – as mentioned above – is playing an increasing role in spruce budworm control, in Maine at least. Inevitably, with increasing use, questions will arise about its effect on non-target biota, in which case the experience gained in laboratory and field evaluation methods will be a great asset in future research.

In addition, unforeseen contamination problems due to chemicals other than the well-established pesticides may arise in future, in which case the impressive range of established methods could be put to use with the minimum of delay. A good example of one of these unforeseen events is provided by the moth-proofing agent Eulan WA New which has largely replaced dieldrin (DLN) in the textile industry in the UK as part of the phasing out of the environmentally undesirable organochlorine insecticides. The course of events in one of these textile areas, Loch Leven, an important sports fishing and recreational area in Scotland, is particularly illuminating (Wells & Cowan, 1984). In a local textile mill, a dieldrin formulation Dielmoth was orginally used. Effluent from the mill discharged into the main tributary stream of the loch, and consequently in the early 1960s high concentrations of dieldrin were found in the loch. In 1964, Dielmoth was withdrawn and replaced by the new compound Eulan WA New which it was assumed would reduce contamination. However, because this compound is a less effective insecticide than DLN, more is required to produce the desired moth proofing. As a result, high concentrations (up to 1800 ng/l) were recorded at the mouth of one of these rivers which contribute 25% of the inflow into the loch. Later it was found that concentrations of Eulan in the three main species of freshwater fish in the loch were actually found to have exceeded previous levels recorded for DLN, indicating a quite unplanned and unpredictable *increase* in contamination of the fresh waters. As the aquatic toxicity of Eulan appears to be not unlike that of DLN, it seems almost certain that the aquatic fauna, macroinvertebrates in particular, must also have been affected by this change.

The conventional sampling methods widely used in field studies have despite their imperfections, been successful in most cases in demonstrating the immediate and short-term effects of pesticides on stream macroinvertebrates. They have also demonstrated fairly consistently that with a wide range of pesticide chemicals used in the post-DDT era, even the most 'catastrophic' impact is followed sooner or later by return of macroinvertebrate populations to normality, with even the most sensitive organisms eventually recovering within a matter of months, or a year at most.

What is proving rather more difficult to demonstrate by conventional methods is the long-term response of invertebrate populations, and of the ecosystem as a whole, to regular or repeated treatments over a period of years. Even in one of the most comprehensively monitored programmes, viz., the OCP in West Africa, opinions about the ecological effects on non-targets after 10 years are still divided. The main difficulties in such long-term evaluation are fully recognised, namely that over a long period faunal changes and fluctuations can occur as a result of many other factors other than toxic chemicals. Nor can much reliance be placed on comparisons between treated rivers and untreated 'controls', even though they may appear superficially to be sufficiently similar for valid comparison.

This has been well demonstrated in faunal studies on two apparently identical streams in southwestern Ontario (Dance & Hynes, 1980) draining adjacent basins. Over the years, differences in agricultural land use brought about physical and chemical changes in the two streams, which in turn produced faunal changes. One stream was particularly affected by cutting forests, which had resulted in intermittent flow and a reduction of species unable to survive periods of no-flow. The other stream, with its source area still surrounded by trees, flowed all the year round. These basic differences in turn produced differences in water chemistry, temperature, and degree of exposure to silt and barnyard run-off. In addition to the differences between the two streams, both were found to support a less diverse fauna than unmodified streams of similar size, unexposed to the intensive agricultural land use in that area.

This factor of intensified land use is one which sooner or later may provide one more complicating factor in long term evaluation in the West African OCP; resettlement of onchocerciasis-free areas being one of the main long-term objectives of the programme. Ideally, some allowance for these natural periodic fluctuations and variations in macroinvertebrate populations could be made if long-term evaluation programmes were based on several years pre-treatment observation on the natural uncontaminated stream or river. This is rarely possible in reality, and indeed, many of the rivers in the West African OCP for example only came into the monitoring programme after river treatment with *Simulium* larvicide had started.

An equally challenging problem in the evaluation of pesticide impact is the ecological effect of long, more or less continuous exposures to very low concentrations of toxic chemical normally considered to be at 'sublethal' levels. There seems to be great scope for inventive research in this field. In one direction would be the use of controlled laboratory experiments with macroinvertebrate species well suited for maintenance in healthy laboratory conditions for long periods, a subject which itself deserves much more attention from freshwater biologists. Another direction would be to study the reactions of sensitive macroinvertebrate indicator species to samples of water from contaminated rivers which have been concentrated by factors of several hundred times. The value of this concentration technique – using fish (*Poecilia reticulatus*) as biological indicators of hazardous chemicals – has been well demonstrated in the Rhine and the Meuse (Sloof *et al.*, 1983b), and there would appear to be great potential for using appropriate macroinvertebrates, whose tolerance levels to a range of chemicals had already been determined, to detect that fraction of the river contamination due to pesticide chemical.

Evaluation of pesticide impact will no doubt be further complicated in some cases by factors which have recently received increasing attention. One of these factors is the capacity of silt to retain residues of pesticide long after they have disappeared from the main water body. This is a familiar feature of contamination by DDT and other persistent chlorinated hydrocarbon insecticides, but recent studies have shown that some of the more modern, non-persistent, pesticides may show a somewhat similar reaction. For example, permethrin, which disappears very quickly from the water of streams and rivers when treatment ceases, can remain in biologically significant concentrations in silt when they are no longer detectable in the water (Friesen, Galloway & Flannagan, 1983). Silt-dwelling fauna, as typified by the burrowing mayfly nymph of *Hexagenia*, and which ingest sediments, are particularly affected. Under experimental conditions it was shown that such sediments in water exposed to normal field dosage rates of permethrin can remain lethal to introduced or 'recolonising' organisms for periods up to 8 days after the supernatant contaminated water has been completely replaced by clean water. Experimental field trials with permethrin have shown that benthic invertebrates may fail to recolonise sections of the stream for up to 6 weeks after treatment (Kingsbury & Kreutzweiser, 1979), long after permethrin residues are no longer detectable in the stream water itself. It appears that the critical factor is not just the retention of chemical by silt and sediments, but the fact that these residues are present in biologically significant concentrations.

Another complicating factor in evaluation which may become increasingly important in running waters exposed to industrial contamination, is the effect on macroinvertebrates of heavy metals which may be present along with pesticide. Macroinvertebrates in general have long been considered much more tolerant to such contamination than are freshwater fishes, but precise information on this point has been scanty. Recently the question has been more critically examined using such typical stream invertebrates as *Ptero-narcys*, *Hydropschye*, *Brachycentrus*, *Ephemerella* and *Gammarus*, with particular reference to cadmium and lead toxicity (Spehar, Anderson & Fiandt, 1978).

Experiments based on 28 day exposures showed that the susceptibility levels of some test species were much the same as fish exposed over their complete cycle in water of similar quality. In addition, test invertebrates were able to accumulate cadmium and lead concentrations up to 30 000 times and 9000 times greater than the corresponding metal concentrations in the water. It is clear that when pesticide chemicals are accompanied by other contaminants of this nature, the question of attributing faunal changes to any one particular contaminant may become extremely complex.

REFERENCES

AAB (Association of Applied Biologists, UK) (1981). Aquatic weeds and their control. National Vegetable Res. Stn, Wellesbourne, Warwick, England.

Abban, E. K. & Samman, J. (1980). Preliminary observations on the effect of the insect larvicide Abate on fish catches in the River Oti, Ghana. *Environmental Pollution*, **21**, 301–11.

Abel, P. D. (1980). Toxicity of hexachlorocyclohexane (Lindane) to *Gammarus pulex*; mortality in relation to concentration and duration of exposure. *Freshwater Biology*, **10**, 251–9.

Abram, F. S. H. (1973). Apparatus for control of poison concentration in toxicity studies with fish. *Water Research*, **7**, 1875–9.

Adams, W. J., Kimerle, R. A., Heidolph, B. B. & Michael, P. R. (1983). Field comparison of laboratory-derived acute and chronic toxicity data. Aquatic Toxicology and Hazard Assessment, Sixth Symposium, *American Society for Testing and Materials, Tech. Publ.*, ed. W. E. Bishop, R. D. Caldwell & B. B. Heidolph, **802**, pp. 367–85. Philadelphia: American Society for Testing and Materials.

Alabaster, J. S. (1969). Survival of fish in 164 herbicides, insecticides, fungicides, wetting agents and miscellaneous substances. *International Pest Control*, **11**, 29–35.

Alabaster, J. S. & Abram, F. S. H. (1965). Estimating the toxicity of pesticides to fish. *P.A.N.S.*, **11**, 91–7.

Alabaster, J. S. & Lloyd, R. (1980). *Water Quality Criteria for Freshwater Fish*. London: Butterworths.

Allen, J. D. & Russek, E. (1985). The quantification of stream drift. *Canadian Journal of Fisheries and Aquatic Sciences*, **42**, 210–15.

Ali, A. & Mulla, M. S. (1977). The IGR diflubenzuron and organophosphorus insecticides against nuisance midges in man-made residential-recreational lakes. *Journal of Economic Entomology*, **70**(5), 571–7.

Ali, A. & Mulla, M. S. (1978). Effects of chironomid larvicides and diflubenzuron on non-target invertebrates in residential recreational lakes. *Environmental Entomology*, **7**(1), 21–7.

Aly, O. A. & Badawy, M. I. (1984). Organochlorine residues in fish from the River Nile, Egypt. *Bulletin of Environmental Contamination and Toxicology*, **33**, 246–52.

American Chemical Society (1972). Fate of organic pesticides in the aquatic environment. *American Chemical Society Symposium, Advances in Chemistry*, Series III, ed. R. F. Gould. Washington DC: American Chemical Society.

American Public Health Association (1976). *Standard Methods for the Examination of Water and Wastewater*, 14th edn. Washington: Published by APHA, American Waterworks Association, and American Water Pollution Control Federation. (Latest, 16th edn. published 1985.)

American Society for Testing and Materials (1982). *Aquatic Toxicity and Hazard Assessment*, Fifth Conference, ASTM STP 766, ed. J. G. Pearson, R. B. Foster & W. E. Bishop. Philadelphia.

243

References

American Society for Testing and Materials (1983). *Aquatic Toxicity and Hazard Assessment.* Sixth Symposium, ASTM STP 802, eds. W. E. Bishop, R. D. Caldwell & B. B. Heidolph. Philadelphia.

Anderson, N. H. & Sedell, J. R. (1979). Detritus processing by macroinvertebrates in stream ecosystems. *Annual Review of Entomology*, **24**, 351–77.

Anderson, R. L. (1982). Toxicity of fenvalerate and permethrin to several non-target aquatic invertebrates. *Environmental Entomology*, **11**, 1251–7.

Anderson, R. L. & DeFoe, D. L. (1980). Toxicity and bioaccumulation of endrin and methoxychlor in aquatic invertebrates and fish. *Environmental Pollution (A)*, **22**, 111–21.

Anderson, R. L. & Shubat, P. (1984). Toxicity of flucythrinate to *Gammarus lacustris* (Amphipoda), *Pteronarcys dorsata* (Plecoptera) and *Brachycentrus americanus* (Trichoptera); importance of exposure duration. *Environmental Pollution (A)*, **35**, 353–65.

Andrews, M., Flower, L. S., Johnstone, D. R. & Turner, C. R. (1983). Spray droplet assessment and insecticide drift studies during the large scale aerial application of endosulphan to control *Glossina morsitans* in Botswana. *Tropical Pest Management*, **29**(3), 239–48.

Applegate, V. C., Howell, J. H., Moffett, J. W., Johnson, B. G. H. & Smith, M. A. (1961). Use of 3-trifluoromethyl-4-nitrophenol as a selective sea lamprey larvicide. *Great Lakes Fishery Commission, Tech. Rep.* No. 1. 35 pp.

Arthur, J. W. (1980). Review of freshwater bioassay procedures for selected amphipods. Symposium on aquatic invertebrate bioassays. *American Society for Testing and Materials. Tech. Publ.* ed. A. L. Buikema & J. Cairns, 715, pp. 98–108. Philadelphia: American Society for Testing and Materials.

Arthur, J. W., Lemke, A. E., Mattson, V. R. & Halligan, B. J. (1974). Toxicity of sodium nitrilotriacetate to the fathead minnow and an amphipod in soft water. *Water Research*, **8**, 187–93.

Arthur, J. W., Zischke, J. A., Allen, K. N. & Hermanutz, R. O. (1983). Effects of diazinon on macroinvertebrates and insect emergence in outdoor experimental channels. *Aquatic Toxicology*, **4**, 283–301.

Baldry, D. A. T., Everts, J., Roman, B., von Ochssee, G. A. B. & Laveissiere, C. (1981). The experimental application of insecticides from a helicopter for the control of riverine populations of *Glossina tachinoides* in West Africa. Part VIII: The effects of two spray applications of OMS-570 (endosulphan) and of OMS-1998 (decamethrin) on *G. tachinoides* and non-target organisms in Upper Volta. *Tropical Pest Management*, **27**(1) 83–110.

Baldry, D. A. T., Kulzer, H., Bauer, S., Lee, C. W. & Parker, J. D. (1978a). The experimental application of insecticides from a helicopter for the control of riverine populations of *Glossina tachinoides* in West Africa. III. Operational aspects and application techniques. *Pesticide Abstracts and News Summary*, **24**(4), 423–34.

Baldry, D. A. T., Molyneux, D. H. & van Wettere, P. (1978b). The experimental application of insecticides from a helicopter for the control of riverine populations of *Glossina tachinoides* in West Agrica. V. Evaluation of decamethrin applied as a spray. *Pesticides Abstracts and News Summary*, **24**(4), 447–54.

Balk, I. F. & Koeman, J. H. (1984). Future hazards from pesticide use. *International Union for Conservation of Nature and Natural Resources* (IUCN). *Commission on Ecology*, Paper No. 4, 100 pp.

References

Bass, J. A. B., Ladle, M. & Welton, J. S. (1982). Larval development and production by the net-spinning caddis, *Polycentropus flavomaculatus*. Pictet (Trichoptera) in a recirculating stream channel. *Aquatic Insects*, **4**, 137–51.

Bays, L. R. (1969). Pesticide pollution and the effects on the biota of Chew Valley Lake. *Journal of the Society for Water Treatment and Examination*, **18**, 295–326.

Bedford, J. W., Raolfs, E. W. & Zabik, M. J. (1968). The freshwater mussel as a biological monitor of pesticide concentration in a lotic environment. *Limnology and Oceanography*, **13**, 118–26.

Bedford, J. W. & Zabik, M. J. (1973). Bioactive compounds in the aquatic environment: uptake and loss of DDT and dieldrin by freshwater mussels. *Archives of Environmental Contamination and Toxicology*, **1**(2) 97–111.

Benfield, E. F. & Buikema, A. I. (1980). Synthesis of miscellaneous invertebrate toxicity tests. Symposium on aquatic invertebrate bioassays. *American Society for Testing and Materials*, *Tech. Publ.*, ed. A. L. Buikema & J. Cairns, 715, pp. 174–87. Philadelphia: American Society for Testing and Materials.

Berry, C. R., Schreck, C. B. & Van Horn, S. L. (1975). Aquatic macroinvertebrate response to field application of the combined herbicides diquat and endothal. *Bulletin of Environmental Contamination and Toxicology*, **14**(3), 374–9.

Birmingham, B. C. & Colman, B. (1985). Persistence and fate of 2,4-D butoxyethanol ester in artificial ponds. *Journal of Environmental Quality*, **14**(1), 100–4.

Birmingham, B. C., Thorndyke, M. & Colman. (1983). The dynamics and persistence of the herbicide Aquakleen in small artificial ponds and its impact on non-target aquatic microflora and microfauna. *Canadian Technical Reports on fisheries and Aquatic Science*, **1151**, 12–23. Ottawa: Dept. of Fish & Oceans.

Bishop, J. E. & Hynes, H. B. N. (1969). Downstream drift of the invertebrate fauna in a stream ecosystem. *Archives für Hydrobiologie*, **66**, 56–90.

Bluzat, R. & Senge, J. (1979). Effets de trois insecticides (lindane, fenitrothion et carbaryl): toxicité aigué sur quartre espèces d'invertébrés. *Environmental Pollution*, **18**, 51–70.

Boryslawskyj, M. & Garood, A. C. (1985). Spatial and temporal patterns of dieldrin pollution in the Holme catchment, West Yorkshire, England. *Environmental Pollution (B)*, **10**, 129–39.

Bowen, H. M. (1982). A survey of sheep dip usage in Scotland in 1978. *Pesticide Science*, **13**, 563–74.

Boyle, T. P. (1980). Effects of the aquatic herbicide 2,4-D DMA on the ecology of experimental ponds. *Environmental Pollution A* **21**(1), 35–49.

Brenner, R. J. & Cupp, E. W. (1980). Rearing blackflies (Diptera: Simuliidae) in a closed system of water circulation. *Tropenmedizin und Parasitologie*, **31**, 247–58.

Brooker, M. P. (1976). The ecological effects of the use of Dalapon and 2,4-D for drainage channel management. I. Flora and chemistry. *Archives für Hydrobiologie*, **78**(3), 396–412.

Brooker, M. P. & Edwards, R. W. (1974). Effects of the herbicide paraquat on the ecology of a reservoir. III. Fauna and general discussion. *Freshwater Biology*, **14**, 311–35.

Brooker, M. P. & Morris, D. L. (1980). A survey of the macroinvertebrate riffle fauna of the River Wye. *Freshwater Biology*, **10**, 437–58.

Brown, L. G., Bellinger, E. G. & Day, J. P. (1979). Dieldrin pollution in Holme catchment, Yorkshire. *Environmental Pollution*, **18**, 203–11.

245

References

Brown, V. M. (1973). Concepts and outlook in testing the toxicity of substances to fish. In *Bioassay Techniques and Environmental Chemistry*. ed. G. E. Glass, pp. 73–95. Ann Arbor: Ann Arbor Science Publishers.

Brown, V. M. (1976). Advances in testing the toxicity of substances to fish. *Chemistry and Industry*, 4, 143–9.

Brungs, W. A. & Mount, D. I. (1970). A water delivery system for small fish-holding tanks. *Transactions of the American Fisheries Society*, 99(4), 799–802.

Buikema, A. L. & Cairns, J. (eds) (1980). Aquatic invertebrate bioassays. *American Society for Testing Materials, Technical Publication*, 715, 209 pp. Philadelphia: American Society for Testing Materials.

Buikema, A. L., Geiger, J. G. & Lee, D. R. (1980). Daphnia toxicity tests. Symposium on aquatic invertebrate bioassays. *American Society for Testing and Materials, Tech. Publ.*, ed. A. L. Buikema & J. Cairns, 715, pp. 48–69.

Burdick, G. E., Dean, H. J., Harris, E. J., Skea, J., Frisa, C. & Sweeney, C. (1968). Methoxychlor as a blackfly larvicide, persistence of its residue in fish and its effect on stream arthropods. *New York Fish Game J.*, 15(2), 121–42.

Burke, W. D. & Ferguson, D. E. (1968). A simplified through-flow apparatus for maintaining fixed concentrations of toxicants in water. *Transactions of the American Fisheries Society*, 97(4), 498–501.

Burress, R. M. & Luhning, C. W. (1969a). Field trials of antimycin as a selective toxicant in channel catfish ponds. *Investigations in Fish Control*, No. 25. Washington: US Dept of the Interior.

Burress, R. M. & Luhning, C. W. (1969b). Use of antimycin for selective thinning of sunfish populations in ponds. *Investigations in Fish Control*, No. 28. Washington: US Dept of the Interior.

Burton, W. & Flannagan, J. F. (1976). An improved river drift sampler. *Fisheries and Marine Services, Environment, Canada, Tech. Rep.* No. 641.

Butijn, G. D. & Koeman, J. H. (1977). Evaluation of the possible impact of aldrin, dieldrin and endrin on the aquatic environment. Brussels, Commission of the European Communities. *Environment and Consumer Protection Service*, ENV/471/77.

Chadwick, G. G. & Brocksen, R. W. (1969). Accumulation of dieldrin by fish and selected fish-food organisms. *Journal of Wildlife Management*, 33(3), 693–700.

Challier, A., Laveissiere, C., Eyraud, M., Kulzer, H., Pawlick, O. & Krupke, M. (1974). Helicopter application of insecticide to control riverine tsetse flies in the West African savanna. *World Health Organization Documentary Series*. WHO/VBC/74. 27 p.

Chance, M. M. (1970). A review of chemical control methods for blackfly larvae (Diptera: Simuliidae). *Quaestiones Entomologicae*, 6, 287–92.

Chandler, J. H. & Marking, L. L. (1975). Toxicity of the lampricide 3-trifluoromethyl-4-nitrophenol (TFM) to selected aquatic invertebrates and frog larvae. *Investigations in Fish Control*, No. 62. Washington: US Dept of the Interior.

Chandler, J. H., Sanders, H. O. & Walsh, D. F. (1974). An improved chemical delivery apparatus for use in intermittent-flow bioassays. *Bulletin of Environmental Contamination and Toxicology*, 12(1), 123–8.

Chapman, G. A. (1983). Do organisms in laboratory toxicity tests respond like organisms in nature? Aquatic Toxicology and Hazard Assessment; Sixth Symposium. *American Society for Testing and Materials, Tech Publ.* ed. W. E. Bishop, R. D. Caldwell,

References

B. B. Heidolph, 802, pp. 315–27. Philadelphia: American Society for Testing and Materials.

Charnetski, W. A., Depner, K. R. & Beltaos, S. (1980). Distribution and persistence of methocychlor in Athabasca river water. In *Control of Blackflies in the Athabasca River, Technical Report, Alberta Environment*, ed. W. O. Haufe & G. C. R. Croome pp. 39–61. Edmonton: Pesticides Chemical Branch, Pollution Control Division.

Chaston, I. (1968). Endogenous activity as a factor in invertebrate drift. *Archives für Hydrobiologie*, **64**(3), 324–34.

Cole, H., Barry, D. & Frear, D. E. H. (1967). DDT levels in fish, stream sediments and soil before and after aerial spraying applications for Fall Cankerworm in northern Pennsylvania. *Bulletin of Environmental Contamination and Toxicology*, **2**(3), 127–46.

Coleman, M. H. & Hynes, H. B. N. (1970). The vertical distribution of the invertebrate fauna in the bed of a stream. *Limnology and Oceanography*, **15**, 31–40.

Cope, O. B. (1966). Contamination of the freshwater ecosystem by pesticides. *Journal of Applied Ecology*, **3** (Suppl) 33–44.

Corbet, S. A., Green, J. & Betney, E. (1973). A study of a small tropical lake treated with the molluscicide Frescon. *Environmental Pollution*, **4**(3), 193–206.

Corbet, P. S., Schmid, F. & Augustin, C. L. (1966). The Trichoptera of St Helen's Island, Montreal. I. The species present and their relative abundance at light. *Canad. Ent.*, **98**, 1284–98.

Courtemanch, D. L. & Gibbs, K. E. (1980). Short- and long-term effects of forest spraying of carbaryl (Sevin-4-oil) on stream invertebrates. *Canadian Entomologist*, **112**, 271–6.

Cowell, B. C. & Carew, W. C. (1976). Seasonal and diel periodicity in the drift of aquatic insects in a subtropical Florida stream. *Freshwater Biology*, **6**, 587–94.

Craig, D. A. (1966). Techniques for rearing stream-dwelling organisms in the laboratory. *Tuatara*, **14**(2), 65–72.

Crossland, N. O. (1982). Aquatic toxicology of cypermethrin. II. Fate and biological effects of pond experiments. *Aquatic Toxicology*, **2**, 205–22.

Crossland, N. O. (1984). Fate and biological effects of methyl parathion in outdoor ponds and laboratory aquaria. II. Effects. *Ecotoxicology & Environmental Safety*, **8**, 482–95.

Crossland, N. O. & Bennett, D. (1984). Fate and biological effects of methyl parathion in outdoor ponds and laboratory aquaria. I. Fate. *Ecotoxicology & Environmental Safety*, **8**, 471–81.

Crossland, N. O., Shires, S. W. & Bennett, D. (1982). Aquatic toxicology of cypermethrin. III. Fate and biological effects of spray drift deposits in freshwater adjacent to agricultural land. *Aquatic Toxicology*, **2**, 253–70.

Cummins, K. W. (1973). Trophic relations of aquatic insects. *Annual Review of Entomology*, **18**, 183–206.

Cummins, K. W. (1974). Structure and function of stream ecosystems. *Biological Science*, **24**, 631–41.

Cummins, K. W. & Klug, M. J. (1979). Feeding ecology of stream invertebrates. *Annual Review of Ecological Systems*, **10**, 147–72.

Cummins, K. W., Petersen, R. C., Howard, F. O., Wuycheck, J. C. & Holt, V. I. (1983). The utilization of leaf litter by stream detritivores. *Ecology*, **54**, 336–45.

Dance, K. W. & Hynes, H. B. N. (1980). Some effects of agricultural land use on stream insect communities. *Environmental Pollution (A)*, **19**, 19–28.

References

Davies, H. (1971). Further eradication of tsetse in the Chad and Gongola River systems in north-eastern Nigeria. *Journal of Applied Ecology*, **8**, 563–78.

Davies, J. B., Gboho, C., Baldry, D. A. T., Bellec, C., Sawadogo, R. & Tiao, P. C. (1982). The effects of helicopter applied adulticides for riverine tsetse control of *Simulium* populations in a West African savanna habitat. I. Introduction, methods and the effect on biting adults and aquatic stages of *Simulium damnosum* s.l. *Tropical Pest Management*, **28**(3), 284–90.

Davies, J. B., Le Berre, R. Walsh, J. F. & Cliff, B. (1978). Onchocerciasis and *Simulium* control in the Volta River Basin. *Mosquito News*, **38**(4), 466–72.

Davies, J. E. & Bowles, J. (1979). Effect of large scale aerial applications of endosulphan on tsetse fly, *Glossina morsitans centralis machado*, in Botswana. *Centre for Overseas Pest Research (COPR) Miss. Rep.* **45**.

DeFoe, D. L. (1975). Multi-channel toxicant injection system for flowthrough bio-assays. *Journal of the Fisheries Research Board of Canada*, **32**, 544–6.

Dejoux, C. (1975). Nouvelle technique pour tester *in situ* l'impact de pesticides sur la faune aquatique non cible. *Cah. ORSTOM. ser. ent. med. et. Parasit.*, **13**(2), 75–80.

Dejoux, C. (1978). Traitements des ecosystemes lotique tropicaux aux insecticides organophosphores. *Verhandlung der Internationalen Vereinigung für Theoretische und Angewandte Limnologie*, **20**, 2708–13.

Dejoux, C. (1982). Recherche sur le devenir des invertébrés derivant dans un cours d'eau tropical a la suite de traitements antisimulidiens au temephos. *Revue Français de Sciences de l'Eau*, **1**, 267–83.

Dejoux, C. & Elouard, J.-M. (1977). Action de l'Abate sur les invertébrés aquatique cinétique de décrochement à court et moyen terme. *Cahiers de l'Office de la Recherche Scientifique et Technique Outre-Mer, Series Hydrobiologie*, **11**(3), 217–30.

Dejoux, C., Elouard, J. M., Justin, J. M., Gibon, F. M. & Troubat, J. J. (1980). Action du temephos (Abate) sur les invertébrés aquatique. VIII, Mise en évidence d'un impact a long terme après six années de surveillance. *Rapport O.R.S.T.O.M.*, No. 36, Bouake, Ivory Coast.

Dejoux, C., Gibon, F. M., Lardeux, F. & Ouattara, A. (1982). Estimation de l'impact du traitement au chlorphoxim de quelques rivières de Côte d'Ivoire durant la saison des pluies 1981. *ORSTOM Report*, No. 47.

Dejoux, C. & Guillet, P. (1980). Evaluation of new blackfly larvicides for use in onchocerciasis control in West Africa. *World Health Organization (Geneva) Documentary series*, WHO/VBC/80.783.

Depner, K. R., Charnetski, W. A. & Haufe, W. O. (1980). Effect of methoxychlor on resident populations of the invertebrates of the Athabasca River. In *Control of blackflies in the Athabasca River, Technical Report*. Alberta Environment, ed. W. O. Haufe, & G. C. R. Croome, pp. 141–50. Edmonton: Pesticides Chemical Branch, Pollution Control Division.

Dermott, R. M. & Spence, H. J. (1984). Changes in populations and drift of stream invertebrates following lampricide treatment. *Canadian Journal of Fisheries and Aquatic Science*, **14**, 1695–701.

Derr, S. K. (1974). Bioactive compounds in the aquatic environment. Loss of methoxychlor from autumn-shed leaves into the aquatic environment. *Bulletin of Environmental Contamination and Toxicology*, **11**(6), 500–2.

Dickson, K. L., Cairns, J. & Arnold, J. C. (1971). An evaluation of the use of a

248

basket-type artificial substrate for sampling macroinvertebrate organisms. *Transactions of the American Fisheries Society*, 100(3), 553–9.

Dimond, J. B. & Malcolm, S. A. (1971). Accothion and aquatic insects: monitoring of stream populations, in *State of Maine* (1971), ed. R. W. Nash, J. W. Peterson & J. F. Chansler, pp. 60–8. Augusta, Maine: State of Maine, US Department of Agriculture & US Department of the Interior.

Doby, J. M., Rault, B. & Beaucournu-Saguez, F. (1967). Utilisation de rubans de plastique pour la récolte des oeufs et des stades larvaires et nymphaux de Simulies (Dipteres Paranematoceres) et pour l'étude biologique de ceux-ci. *Annales de Parasitologie, Paris*), 42(6), 651–7.

Douben, P. E. T., Everts, J. W. & Koeman, J. H. (1985). *Side Effects of Pyrethroids Used to Control Adults of Simulium damnosum s.l. in a Riverine Forest in Togo (West Africa)*. Wageningen, Netherlands: Agricultural University

Douthwaite, R. J., Fox, P. J., Matthiessen, P. & Russell-Smith, A. (1981). The environmental impact of aerosols of endosulphan applied for tsetse fly control in the Okavango Delta, Botswana. *Final Report of the Endosulphan Monitoring Project*. London: Overseas Development Administration.

Drake, C. M. & Elliott, J. M. (1983). A new quantitative air-life sampler for collecting macroinvertebrates on stony bottoms in deep rivers. *Freshwater Biology*, 13, 545–9.

Duncan, J., Brown, N. & Dunlop, R. W. (1977). The uptake of the molluscicide 4'-chloronicotinanilide into *Biomphalaria glabrata* (Say) in a flowing water system. *Pesticide Science*, 8, 345–53.

Duthie, J. R. (1977). The importance of sequential assessment in test programmes for estimating hazard to aquatic life. In *Aquatic toxicology and hazard evaluation*, ed. F. L. Mayer & J. L. Hamelink. Philadelphia: American Society for Testing Materials. Special Technical Publication 634.

Eco-Analysis Inc. (1982). The monitoring of carbaryl at river mouths. *Env. Monitor. Rep.*, 1981, pp. 123–31. Maine cooperative spruce budworm suppression project. Augusta, Maine: Dept. of Conservation.

EEC (European Economic Commission) (1981). Water quality criteria for European freshwater fish. Report on the effects produced by combinations of toxicants in the water on freshwater fish and other aquatic fauna. *Food and Agricultural Organization, Rome, Tech. Doc.* CECPI/T37.

EEC (1984). Water Quality criteria for European freshwater fish. Report on nitrite and freshwater fish. *EIFAC Tech. Rep.*, 46.

Eichelberger, J. W. & Lichtenberg, J. J. (1971). Persistence of pesticides in river water. *Environmental Science and Technology*, 5, 541–4.

Eidt, D. C. (1975). The effect of fenitrothion from large-scale forest spraying on benthos in New Brunswick headwater streams. *Canadian Entomologist*, 107, 743–60.

Eidt, D. C. (1977). Effects of fenitrothion on benthos in the Nashwaak project study streams in 1976. *Info. Rep. M–X-70*. Fredericton, New Brunswick: Maritimes Forest Research Centre.

Eidt, D. C. (1978). Toxicity of fenitrothion in insects in a woodland stream. *Info. Rep. M–X-86*. Fredericton, New Brunswick: Maritimes Forest Research Centre.

Eidt, D. C. (1981). Recovery of aquatic arthropod populations in a woodland stream after depletion by fenitrothion treatment. *Canadian Entomologist*, 113, 303–13.

References

Eidt, D. C. & Sundaram, K. M. S. (1975). The insecticide fenitrothion in headwater streams from large-scale forest spraying. *Canadian Entomologist*, 107, 735–42.

Eidt, D. C. & Weaver, C. A. A. (1982). Stream invertebrate drift studies. In *Permethrin in New Brunswick Salmon Nursery Streams*, ed. P. J. Kingsbury, pp. 29–37. Sault Ste. Marie, Ontario: Forest Pest Management Institute.

Eidt, D. C. & Weaver, C. A. A. (1984). Influence of site and fenitrothion contamination of vertical distribution of drift arthropods in a woodland stream. *Canadian Entomologist*, 116, 1425–30.

Eidt, D. C., Weaver, C. A. A. & Kreutzweiser, D. (1983). Stream bottom fauna studies. In *Permethrin in New Brunswick Salmon Nursery Streams*, ed., P. D. Kingsbury, pp. 38–50. Sault Ste. Marie, Ontario: Forest Pest Management Institute.

Elliott, J. M. (1965). Daily fluctuations of drift invertebrates in a Dartmoor stream. *Nature*, 205(4976), 1127–9.

Elliott, J. M. (1967a). Invertebrate drift in a Dartmoor stream. *Archives für Hydrobiologie*, 63(2), 202–37.

Elliott, J. M. (1967b). The life histories and drifting of the Plecoptera and Ephemeroptera in a Dartmoor stream. *Journal of Animal Ecology*, 36, 343–62.

Elliott, J. M. (1968a). The life histories and drifting of Trichoptera in a Dartmoor stream. *Journal of Animal Ecology*, 37, 615–25.

Elliott, J. M. (1968b). The daily activity patterns of mayfly nymphs (Ephemeroptera). *Journal of Zoology, London*, 155, 201–21.

Elliott, J. M. (1970). Methods of sampling invertebrate drift in running water. *Annales de Limnologie*, 6(2), 133–59.

Elliott, J. M. (1971a). Some methods for the statistical analysis of samples of benthic invertebrates. *Freshwater Biol. Assn. Sci. Publ*, No. 25.

Elliott, J. M. (1971b) Upstream movements of benthic invertebrates in a Lake District stream. *Journal of Animal Ecology*, 40, 235–52.

Elliott, J. M. (1971c) The distances travelled by drifting invertebrates in a Lake Distict stream. *Oecologia*, 6, 350–79.

Elouard, J. M. (1984). Impact d'un insecticide organophosphore (le temephos) sur les entomocenoses associées aux stades préimaginaux du complexe *Simulium damnosum* Theobald (Diptera. Simuliidae) *O.R.S.T.O.M. Traveux et documents microédites (TDM)*, 13, Thesis to University of Paris-Sud.

Elouard, J. M. & Leveque, C. (1975). Observations préliminaires sur la dérive des invertébrés, et des poissons dans quelques rivières de Côte d'Ivoire. *Rapport Office de la Récherche Scientifique et Technique Outre-Mer*, Bouake No. 75.

Elouard, J. M. & Troubat, H. J. (1979). Actions de l'Abate (temephos) sur les invertébrés aquatique. VII. Effets des premiers traitements de la Basse Maraoue. *Rapport Office de la Récherche Scientifique et Technique Outre-Mer*, Bouake No. 32.

Elson, P. F., Saunders, J. W. & Sitko, V. (1972). Impact of forest-based industries on freshwater dependent fish resources in New Brunswick. *Fisheries Research Board of Canada, Tech. Rep.*, No. 325.

Elzorgani, G. A. (1976). Residues of organochlorine insecticides in some fishes and birds in the Gezira of Sudan. *Pesticide Science*, 7, 150–2.

Elzorgani, G. A., Abdulla, A. M. & Ali, M. E. T. (1979). Residues of organochlorine insecticide in fishes in Lake Nubia. *Bulletin of Environmental Contamination and Toxicology*, 22, 44–8.

Everts, J. W., Frankenhuyzen, H, van., Roman, B. & Koeman, J. H. (1983). Side effects of experimental pyrethroid applications for the control of tsetse flies in a riverine

250

References

forest habitat (Africa). *Archives of Environmental Contamination and Toxicology*, 12, 91–7.

Fairchild, G. B. & Barreda, E. A. (1945). DDT as a larvicide against *Simulium*. *Journal of Economic Entomology*, 38, 694–9.

Feldmeth, C. R. (1970). A large volume laboratory stream. *Hydrobiologia*, 35, 397–400.

Fikes, M. H. & Tubb, R. A. (1972). Dieldrin uptake in the three-ridge naiad. *Journal of Wildlife Management*, 36(3), 802–8.

Flannagan, J. F. (1973). Field and laboratory studies on the effect of exposure to fenitrothion on freshwater aquatic invertebrates. *Manitoba Entomologist*, 7, 15–25.

Flannagan, J. F., Townsend, B. E., de March, B. G. E., Friesen, M. K. & Leonhard, S. L. (1979). The effects of an experimental injection of methoxychlor on aquatic invertebrates: accumulation, standing crop, and drift. *Canadian Entomologist*, 111, 73–89.

Flannagan, J. F., Townsend, B. E., & de March, B. G. E. (1980a) Acute and long term effects of methoxychlor larviciding on the aquatic invertebrates of the Athabasca River, Alberta. In *Control of Blackflies in the Athabsca River, Technical Report*. Alberta Environment, ed. W. O. Haufe & G. C. R. Croome, pp. 151–8. Edmonton: Pesticides Chemical Branch, Pollution Control Division.

Flannagan, J. F., Townsend, B. E., de March, G. E., Friesen, M. K. & Leonhard, S. L. (1980b). Effects of an experimental injection of methoxychlor in 1974 on aquatic invertebrates; accumulation, standing crop, and drift. In *Control of blackflies in the Athabasca River, Technical Report*. Ed. W. O. Haufe & G. C. R. Croome, Alberta Environment, pp. 131–9. Edmonton: Pesticides Chemical Branch, Pollution Control Division.

Fox, P. J. & Matthiessen, P. (1982). Acute toxicity to fish of low-dose aerosol applications of endosulphan to control tsetse fly in the Okavango Delta, Botswana. *Environmental Pollution (A)* 27, 129–42.

Frank, R., Braun, H. E., Holdrinet, M. V. H., Sirons, G. J. & Ripley, B. D. (1982). Agriculture and water quality in the Canadian Great Lakes basin: V. Pesticide use in 11 agricultural watersheds and presence in stream water, 1975–77. *Journal of Environmental Quality*, 11(3), 497–505.

Fredeen, F. J. H. (1962). DDT and heptachlor as black-fly larvicides in clear and turbid water. *Canadian Entomologist*, 94, 875–80.

Fredeen, F. J. H. (1969). A new procedure allowing replicated miniature larvicide tests in a large river. *Canadian Entomologist*, 101(7), 713–25.

Fredeen, F. J. H. (1974). Tests with single injections of methoxychlor black fly (Diptera: Simuliidae) larvicides in large rivers. *Canadian Entomologist*, 106, 285–305.

Fredeen, F. J. H. (1975). Effects of a single injection of methoxychlor black-fly larvicide on insect larvae in a 161-km (100–mile) section of the North Saskatchewan River. *Canadian Entomologist*, 107, 807–17.

Fredeen, F. J. H. (1977). Black fly control and environmental quality with reference to chemical larviciding in Western Canada. *Quaestiones Entomologicae*, 13, 321–25.

Fredeen, F. J. H., Arnason, A. P., Berck, B. & Rempel, J. G. (1953). Further experiments with DDT in the control of *Simulium arcticum* Mall. in the North and South Saskatchewan Rivers. *Canadian Journal of Agricultural Science*, 33, 379–93.

Fredeen, F. J. H., Balba, M. H. & Sasa, J. G. (1975). Residues of methoxychlor and other chlorinated hydrocarbons in water, sand, and selected fauna following

References

injections of methoxychlor black fly larvicide into the Saskatchewan River, 1972. *Pesticide Monitoring Journal*, 8(4), 241–6.

Fremling, C. R. (1975). Acute toxicity of the lampricide 3-trifluoromethyl-4-nitrophenol (TFM) to nymphs of mayflies (Hexagenia sp). *Investigations in Fish Control*, No. 58. Washington: US Dept. of the Interior. Fish & Wildlife Service.

Fremling, C. R. & Mauck, W. L. (1980). Methods for using nymphs of burrowing mayflies (Ephemeroptera, Hexagenia) as toxicity test organisms. Aquatic Invertebrate Bioassays, American Society for Testing and Materials, ed. A. L. Buikema & J. Cairns, pp. 81–97. Philadelphia: American Society for Testing and Materials.

Frempong-Boadu, J. (1966). A laboratory study of the effectiveness of methoxychlor, fenthion and carbaryl against blackfly larvae. (Diptera: Simuliidae) *Mosquito News*, 26(4), 562–4.

Friesen, M. K., Galloway, T. D. & Flannagan, J. F. (1983). Toxicity of the insecticide permethrin in water and sediment to nymphs of the burrowing mayfly *Hexagenia rigida* (Ephemeroptera: Ephemeridae). *Canadian Entomologist*, 115, 1007–14.

Frost, S., Huni, A. & Kershaw, W. E. (1971). Evaluation of a kicking technique for sampling stream bottom fauna. *Canadian Journal of Zoology*, 49(2), 167–73.

Furse, M. T., Wright, J. F., Armitage, P. D. & Moss, D. (1981). An appraisal of pond-net samples for biological monitoring of lotic macroinvertebrates. *Water Research*, 15, 679–89.

Garnham, P. C. C. & McMahon, J. P. (1946). The eradication of *Simulium neavei* Roubaud from an onchocerciasis area in Kenys Colony. *Bulletin of Entomological Research*, 37, 619–28.

Gaufin, A. R., Jensen, L. D. & Nelson, T. (1961). Bioassays determine pesticide toxicity to aquatic invertebrates. *Water and Sewage Works*, 108, 355–9.

Gaufin, A. R., Jensen, L. D., Nebeken, A. V., Nelson, T. & Teel, R. W. (1965). The toxicity of ten organic insecticides to various aquatic invertebrates. *Water and Sewage Works*, 112, 276–9.

Gaugler, R., Molloy, D., Haskins, T. & Rider, G. (1980). A bioassay system for the evaluation of black fly (Diptera: Simuliidae) control agents under simulated stream conditions. *Canadian Entomologist*, 112, 1271–6.

George, J. P. & Hingorani, H. G. (1982). Herbicide toxicity to fish food organisms. *Environmental Pollution (A)*, 28, 183–8.

Gibbs, K. E., Rabeni, C. R., Stanley, J. G. & Trial, J. C. (1979). The effects of a split application of Sevin-4-oil on aquatic organisms. In *Environmental Monitoring of Cooperative Spruce Budworm Control Projects, Maine 1978*. ed. K. G. Stratton, pp. 53–106. Augusta, Maine: Maine Dept. of Conservation, Bureau of Forestry.

Gibbs, K. E., Mingo, T. M., Courtemanch, D. L. & Stairs, D. J. (1981). The effects on pond macroinvertebrates from forest spraying of carbaryl and its persistence in water and sediment. In *Environmental Monitoring Reports from the 1980 Maine Cooperative Spruce Budworm Suppression Project*, ed. K. G. Stratton, pp. 121–47. Augusta, Maine: Maine Forest Service, Dept. of Conservation.

Gibbs, K. E., Mingo, T. M. & Courtemanch, D. L. (1982). The long-term effects on pond macroinvertebrates from forest spraying of carbaryl (Sevin-4-Oil) in 1980 and its persistence in water and sediment in 1981. In *Environmental Monitoring of Cooperative Spruce Budworm Control Project*, ed. K. G. Stratton, pp. 1–20. Augusta Maine: Maine Department of Conservation, Bureau of Forestry.

Gibbs, K. E., Mingo, T. M. & Courtemanch, D. L. (1983). The effects in 1982 on pond macroinvertebrates from forest spraying with carbaryl, (Sevin-4-Oil), in 1980. In Environmental Monitoring of Cooperative Spruce Budworm Control Project, ed. K. G. Stratton. Augusta, Maine: Maine Department of Conservation, Bureau of Forestry.

Gibbs, K. E., Mingo, T. M. & Courtemanch, D. L. (1984). Persistence of carbaryl (Sevin-4-Oil) in woodland ponds and its effect on pond macroinvertebrates following forest spraying. *Canadian Entomologist*, **116**, 203–13.

Gibon, F. M. & Troubat, J. J. (1980). Effets d'un traitement au Chlorphoxim sur la dérive des invertébrés benthique. *Rapport de l'Office de la Récherche Scientifique et Technique Outre-Mer*, No. 27.

Gilderhus, P. A. (1967). Effects of diquat on bluegills and their food organisms. *Progressive Fish Culturist*, **29**, 67–74.

Gilderhus, P. A., Berger, B. L. & Lennon, R. E. (1969). Field trials of antimycin A as a fish toxicant. *Investigations in Fish Control*, No. 27. Washington: US Dept. of the Interior.

Gilderhus, P. A. & Johnson, B. G. H. (1980). Effects of sea lamprey (*Petromyzon marinus*) control in the Great Lakes on aquatic plants, invertebrates and amphibians. *Canadian Journal of Fisheries and Aquatic Science*, **37**, 1895–905.

Goebel, H., Gorbach, S., Knauf, W., Rimpau, R. H. & Huttenbach, H. (1982). Properties, effects, residues, and analytics of the insecticide Endosulphan. *Residue Reviews*, **83**, 1–165.

Gorbach, S., Haaring, R., Knauf, W. & Werner, H. J. (1971*a*). Residue analysis in the water systems of East Java (River Brantas, ponds, seawater) after continued large scale application of Thiodan in rice. *Bulletin of Environmental Contamination and Toxicology*, **6**(1), 40–7.

Gorbach, S., Haaring, R., Knauf, W. & Werner, H. J. (1971*b*). Residue analysis and biotests in rice fields of East Java treated with Thiodan. *Bulletin of Environmental Contamination and Toxicology*, **6**(3), 193–9.

Graham, P. (1964). Destruction of birds and other wildlife by dieldrin spraying against tsetse fly in Bechuanaland. *Arnoldia*, **1**, 1–4.

Green, G. H., Hussain, M. A., Oloffs, P. C. & McKeown, B. A. (1981). Fate and toxicity of acephate (Orthene) added to a coastal B.C. stream. *Journal of Environmental Science and Health*, B, **16**, (3), 253–71.

Hansen, S. R. & Garton, R. R. (1982). Ability of standard toxicity tests to predict the effects of the insecticide Diflubenzuron on laboratory stream communities. *Canadian Journal of Fisheries and Aquatic Science*, **39**, 1273–88.

Hanuska, L. (1971). Airborne pollution of inland water. *Verhandlung der Internationalen Vereinigung für Theoretische und Angewandte Limnologie*, **18**, 968–80.

Harper, D. B., Smith, R. V. & Gotto, D. M. (1977). BHC residues of domestic orihin; a significant factor in pollution of freshwater in Northern Ireland. *Environmental Pollution*, **12**, 223–33.

Harris, C. R. & Miles, J. R. W. (1975). Pesticide residues in the Great Lakes region of Canada. *Residue Reviews*, **57**, 27–79.

Hasegawa, J., Yasuno, M., Saito, K., Nakamura, Y., Hatakeyama, S. & Sato, H. (1982). Impact of temephos and fenitrothion on aquatic invertebrates in a stream of Mt Tsukuba. *Japanese Journal of Sanitary Zoology*, **33**(4), 363–8.

Haufe, W. O. (ed.) (1980). Control of blackflies in the Athabasca River. Evaluation and recommendations for chemical control of *Simulium arcticum* Malloch. Alberta: Pesticide Chemical Branch. Pollution Control Division.

References

Haufe, W. O. & Croome, G. C. R. (eds) (1980). *Control of Black Flies in the Athabasca River*, Technical Report. Edmonton: Pollution Control Division.

Haufe, W. O., Depner, K. R. & Charnetski, W. A. (1980a). Impact of methoxychlor on drifting aquatic invertebrates. In *Control of Black Flies in the Athabasca River, Technical Report*, ed. W. O. Haufe, Alberta Environment, pp. 159–168.

Haufe, W. O., Depner, K. R. & Kozub, G. C. (1980b). Parameters for monitoring displacement of drifting aquatic invertebrates. In *Control of black flies in the Athabasca River, Technical Report*. Alberta Environment, pp. 169–181.

Hellawell, J. M. (1978). *Biological Surveillance of Rivers*. A biological monitoring handbook. Marlow, Bucks: Water Research Centre, Medmenham.

Helson, B. V. & West, A. S. (1978). Particulate formulations of Abate and methoxychlor as black fly larvicides; their selective effects on stream fauna. *Canadian Entomologist*, 110, 591–602.

Herzel, F. (1973). Occurrence of biocides in surface waters illustrated by the example of the Rhine. *Gewasserschutz Wass-abwass*, 10, 367–76.

Hilsenhoff, W. (1966). Effect of diquat on aquatic insects and related animals. *Journal of Economic Entomology*, 59, 1520–1.

Hocking, K. S., Lee, C. W., Beesley, J. S. S. & Matechi, H. T. (1966). Aircraft application of insecticides in East Africa. XVI. Airspray experiment with endosulphan against *Glossina morsitans* Westw., *G. swynnertoni* Aust. and *G. pallidipes* Aust. *Bulletin of Entomological Research*, 56, 737–44.

Holden, A. V. (1972). The effects of pesticides on life in fresh waters. *Proceedings of the Royal Society, London B.*, 180, 383–94.

Holden, A. V. & Bevan, D. (1979). *Control of the Pine Beauty Moth by Fenitrothion in Scotland, 1978*. Forestry Commission. Oxford: Nuffield Press.

Holmes, S. B. & Kingsbury, P. D. (1980). The environmental impact of nonyl phenol and the Matacil formulation. Part I. Aquatic ecosystems. *Forest Pest Management Institute, Canadian Forestry Service, Rep. No. FPM-X-35*. Ontario: Sault Ste. Marie.

Holmes, S. B. & Kingsbury, P. D. (1982). Comparative effects of three Matacil field formulations on stream benthos and fish. *Forest Pest Management Institute, Canadian Forestry Service, Rep. No. FPM-X-55*. Ontario: Sault Ste. Marie.

Holmes, S. B. & Millikin, R. L. (1981). A preliminary report on the effects of a split application of Reldan on aquatic and terrestrial ecosystems. *Forest Pest Management Institute, Canadian Forestry Service*, Rep. No. 13. Ontario: Sault. Ste. Marie.

Holmes, S. B., Millikin, R. L. & Kingsbury, P. D. (1981). Environmental effects of a split application of Sevin-2-oil. *Forest Pest Management Institute, Canadian Forestry Service*, Report No. FPM-X-46. Ontario: Sault Ste. Marie.

Houf, L. J. & Campbell, R. S. (1977). Effects of antimycin A and Rotenone on macrobenthos in ponds. *Investigations in Fish Control*, No. 80. pp. 29. Washington: US Dept. of the Interior, Fish & Wildlife Service.

Howell, J. H., King, E. L., Smith, A. J. & Hanson, L. H. (1964). Synergism of 5,2'-dichloro-4-nitrosalicylanilide and 3-trifluoromethyl-4-nitrophenol in a selective lamprey larvicide. *Great Lakes Fishery Commission Tech, Rep.*, No. 8. Ann Arbor: Great Lakes Fishery Commission.

Howell, J. H. & Marquette, W. M. (1962). Use of mobile bioassay equipment in the chemical control of sea lamprey. *Special Sci. Rep. Fish.*, No. 418. Ann Arbor: US Fish and Wildlife Service.

Huet, M. (1959). Profiles and biology of western European streams as related to fish management. *Transactions of the American Fisheries Society*, 88, 155–63.

References

Hughes, D. A. (1970). Some factors affecting drift and upstream movements of *Gammarus pulex. Ecology*, **51**, 301–5.

Hulbert, P. J. (1978). Mattawamkeag River Studies: II. Effects of Sevin insecticide on fish and invertebrates in 1976. In *Environmental Monitoring of Cooperative Spruce Budworm Control Projects*. Migratory Fish Research Institute, University of Maine, Orono. Augusta: Maine Dept. of Conservation.

Hultin, L. (1968). Upstream movements of *Gammarus pulex* (Amphipoda) in a south Swedish stream. *Oikos*, **22**, 329–47.

Hurlbert, S. H. (1975). Secondary effects of pesticides on aquatic ecosystems. *Residue Reviews*, **57**, 81–148.

Hurlbert, S. H., Mulla, M. S., Keith, K. O., Westlake, W. E. & Dusch, M. E. (1970). Biological effects and persistence of Dursban in freshwater ponds. *Journal of Economic Entomology*, **63** (1), 43–52.

Hydorn, S. B., Rabeni, C. F. & Jennings, D. T. (1979). Effects of forest spraying with acephate insecticide on consumption of spiders by brook trout (*Salvelinus fontinalis*). **111**, 1185–92.

Hynes, H. B. (1961). The effect of sheep dip containing the insecticide BHC on the fauna of a small stream, including *Simulium* and its predators. *Annals of Tropical Medicine and Parasitology*, **55**, 192–96.

Hynes, H. B. N. (1972). *The Ecology of Running Waters*. Liverpool: Liverpool University Press.

Hynes, H. B. N. (1974). Further studies on the distribution of stream animals within the substratum. *Limnology and Oceanography*, **19**, 92–9.

Hynes, H. B. N., Williams, D. D. & Williams, N. E. (1976). Distribution of the benthos within the substratum of a Welsh mountain stream. *Oikos*, **27**, 307–10.

Hynes, J. D. (1975). Downstream drift of invertebrates in a river in Southern Ghana. *Freshwater Biology*, **5**, 515–31.

Jacobi, G. Z. (1971). A quantitative artificial substrate sampler for benthic macro-invertebrates. *Transactions of the American Fisheries Society*, **100**(1), 136–8.

Jacobi, G. Z. & Degan, D. J. (1977). Aquatic macroinvertebrates in a small Wisconsin trout stream before, during, and two years after treatment with the fish toxicant antimycin. *Investigations in Fish Control*, No. 81. Washington: US Dept. of the Interior, Fish & Wildlife Service.

Jamnback, H. (1969). Field tests with larvicides other than DDT for control of blackfly (Diptera: Simuliidae) in New York. *Bulletin of the World Health Organization*, **40**, 635–8.

Jamnback, H. (1976). Simuliidae (Blackflies) and their control. *World Health Organization Documentary Series*, WHO/VBC/76.653.

Jamnback, H. & Frempong-Boadu, J. (1966). Testing blackfly larvicides in the laboratory and in streams. *Bulletin of the World Health Organization*, **34**, 405–21.

Jamnback, H. & Means, R. (1966). Length of exposure period as a factor influencing the effects of larvicides for blackflies. *Mosquito News*, **26**, 589–91.

Jamnback, H. & Means, R. G. (1968). Formulation as a factor influencing the effectiveness of Abate in control of blackflies (Diptera: Simuliidae). *Proceedings of the 55th Annual Meeting of the New Jersey Mosquito Extermination Association*, 89–94.

Jensen, L. D. & Gaufin, A. R. (1964). Effects of ten organic insecticides on two species of stonefly naiads. *Transactions of the American Fisheries Society*, **93**(1), 27–34.

Jensen, L. D. & Gaufin, A. R. (1966). Acute and long term effects of aquatic insecticides on two species of stonefly naiads. *Journal of the Water Pollution Control Federation*, **38**, 1273–86.

Johnson, B. T., Saunders, H. O. & Campbell, R. S. (1971). Biological magnification and degradation of DDT and aldrin by freshwater invertebrates. *Journal of the Fisheries Research Board of Canada*, **28**, 705–09.

Johnson, H. E. & Ball, R. C. (1972). Organic pesticide pollution in an aquatic environment. Fate of organic pesticides in the aquatic environment. *Advances in Chemistry Series*, ed. R. F. Gould, III, pp. 1–10. Washington: American Chemical Society.

Johnson, W. W. & Finley, M. T. (1980). Handbook of acute toxicity of chemicals to fish and aquatic invertebrates. *Resource Publ.*, 137. Washington: US Dept. of the Interior, Fish & Wildlife Service.

Jordan, A. M. (1974). Recent developments in the ecology and methods of control of tsetse flies (*Glossina* supp) (Diptera, Glossinidae) – a review. *Bulletin of Entomological Research*, **63**, 361–99.

Kaushik, N. K. & Hynes, H. B. N. (1968). Experimental study on the role of autumn-shed leaves in aquatic environments. *Journal of Ecology*, **56**, 229–43.

Kawatski, J. A., Dawson, V. K. & Reuvers, M. L. (1974). Effect of TFM and Bayer 73 on *in vivo* oxygen consumption of the aquatic midge *Chironomus tentans*. *Transactions of the American Fisheries Society*, **103**(3), 551–6.

Kawatski, J. A., Ledvina, M. M. & Hansen, C. R. (1975). Acute toxicities of 3-trifluoromethyl-4-nitrophenol (TFM) and 2', 5-dichloro-4-nitrosalicylanilide (Bayer 73) to larvae of the midge *Chironomus tentans*. *Investigations in Fish Control*, No. 57. Washington: US Dept. of the Interior, Fish & Wildlife Service.

Kawatski, J. A. & Zittel, A. E. (1977). Accumulation, elimination, and biotransform-ation of the lampricide 2',5-dichloro-4-nitrosalicylanilide by *Chironomus tentans*. *Investigations in Fish Control*, No. 79. Washington: US Dept. of the Interior, Fish & Wildlife Service.

Kerswill, C. J. (1967). Studies on effects of forest sprayings with insecticides, 1952–63, on fish and aquatic invertebrates in New Brunswick streams: intro-duction and summary. *Journal of the Fisheries Research Board of Canada*, **24**, 701–8.

Kingsbury, P. D. (1976a). Studies on the impact of aerial applications of the synthetic pyrethroid NRDC-143 on aquatic ecosystems. *Info. Rep.* CC-X-127. Ottawa, Ontario: Chemical Control Research Institute.

Kingsbury, P. D. (1976b). Effects of an aerial application of the synthetic pyrethroid permethrin on a forest stream. *Manitoba Entomologist*, **10**, 9–17.

Kingsbury, P. D. (1982). Post spray observations on fish and aquatic invertebrates. In *Permethrin in New Brunswick Salmon Nursery Streams*, ed. P. D. Kingsbury, pp. 69–70. Sault Ste. Marie. Ontario: Forest Pest Management Institute.

Kingsbury, P. D. (ed.) (1983). *Permethrin in New Brunswick Salmon Nursery Streams*, *Info. Rep.* FPM-X-52. Sault Ste. Marie, Ontario: Forest Pest Management Institute.

Kingsbury, P. D. & Holmes, S. B. (1980). A preliminary report on the impact on stream fauna of high dosage applications of Reldan. *File Rep.*, No. 2. Sault Ste. Marie, Ontario: Forest Pest Management Institute.

Kingsbury, P. D., Holmes, S. B. & Millikin, R. L. (1980). Environmental effects of a double application of azamethiphos on selected terrestrial and aquatic organisms. *Rep. No* FPM-X-33. Sault Ste. Marie, Ontario: Forest Pest Management Institute.

Kingsbury, P. D. & Kreutzweiser, D. P. (1979). Impact of double applications of

References

permethrin on forest streams and ponds. *Rep.* FPM-X-27. Sault Ste. Marie, Ontario: Forest Pest Research Institute.

Kingsbury, P. D. & Kreutzweiser, D. P. (1980*a*). Environmental impact assessment of a semi-operational permethrin application. *Rep.* FPM-X-30. Sault Ste. Marie, Ontario: Forest Pest Management Institute.

Kingsbury, P. D. & Kreutzweiser, D. P. (1980*b*). Dosage-effect studies on the impact of permethrin on trout streams. *Rep.* FPM-X-31. Sault Ste. Marie, Ontario: Forest Pest Management Institute.

Koeman, J. H., den Boer, W. M. J., Feith, A. F., de Iongh, H. H., Spliethoff, P. C. (1978). Three years observations on side effects of helicopter applications of insecticides used to exterminate *Glossina* species in Nigeria. *Environmental Pollution*, **15**, 31–59.

Koeman, J. H. & Pennings, J. H. (1970). An orientation survey of the side effects and environmental distribution of insecticides used in tsetse control in Africa. *Bulletin of Environmental Contamination and Toxicology*, **5**, 164–70.

Koeman, J. H., Rijksen, H. D., Smies, M., Na'isa, B. K. & Maclennan, K. J. R. (1971). Faunal changes in a swamp habitat in Nigeria sprayed with insecticide to exterminate Glossina. *Netherlands Journal of Zoology*, **21**(4), 434–63.

Kolton, R. J., MacMahon, P. D., Jeffrey, K. A. & Beamish, F. W. H. (1985). Effects of the Lampricide, 3-Trifluoromethyl-4-nitrophenol (TFM) on the Macroinvertebrates of a Hardwater River. Ontario, Canada: University of Guelph.

Kpetaka, A. E. (1975). Polychlorinated biphenyls (PCB's) in the rivers Avin and Frome. *Bulletin of Environmental Contamination Toxicology*, **14**(6), 687–91.

Kreutzweiser, D. P. (1982). The effects of permethrin on the invertebrate fauna of a Quebec forest. *Rep.* FPM-X-50. Sault. Ste. Marie, Ontario: Forest Pest Management Institute.

Kreutzweiser, D. P. & Kingsbury, P. D. (1982). Recovery of stream benthos and its utilization by native fish following high dosage permethrin applications. *Info. Rep.*, FPM-X-59. Sault Ste. Marie, Ontario: Forest Pest Management Institute.

Lacey, L. A. & Mulla, M. S. (1977*a*). A new bioassay unit for evaluating larvicides against black flies. *Journal of Economic Entomology*, **70**, 453–6.

Lacey, L. A. & Mulla, M. S. (1977*b*). Larvicidal and ovicidal activity of Dimilan against *Simulium vittatum*. *Journal of Economic Entomology*, **70**, 369–73.

Ladle, M., Baker, J. H., Casey, H. & Farr, I. S. (1977). Preliminary results from a recirculating experimental system; observations on interaction between chalk stream water and inorganic sediment. In *Interactions Between Sediments and Freshwater*, ed. H. C. Gollerman, pp. 252–7. The Hague: W. Junk.

Ladle, M., Casey, H., Marker, A. F. H., & Welton, J. S. (1981). The use of large experimental channels for ecological research. Proceedings of a World Symposium on aquaculture in heated effluents and recirculation systems, Stavanger, May 1980, **1**, 279–87. Berlin.

Ladle, M., Welton, J. S. & Bass, J. A. B. (1980). Invertebrate colonization of the gravel substratum of an experimental recirculating channel. *Holarctic Ecology*, **3**, 116–23.

Lauff, G. H. & Cummins, K. W. (1964). A model stream for studies in lotic ecology. *Ecology*, **45**, 188–91.

Lawrence, S. G. (ed.) (1981). Manual for the culture of selected freshwater invertebrates. *Canadian Special Publ. Fisheries & Aquatic Sci.*, **54**. Ottawa: Dept. of Fisheries and Oceans.

Lee, C. W., Parker, J. D., Baldry, D. A. T. & Molyneux, D. H. (1978). The experimental application of insecticides from a helicopter for the control of riverine populations of Glossina tachinoides in West Africa. II. Calibration of equipment and insecticide dispersal. *Pesticides Abstracts and News Summary*, 24(4), 404–22.

Lee, G. F. & Jones, R. A. (1983). Translation of laboratory results to field conditions; the role of aquatic chemistry in assessing toxicity. Aquatic Toxicology and Hazard Assessment; Sixth symposium. *American Society for Testing and Materials, Tech. Publ.* 802, pp. 328–49.

Lennon, R. E. & Berger, B. L. (1970). A resumé of field applications of antimycin A to control fish. *Investigations in Fish Control*, No. 40. Washington: US Dept. of the Interior. Fish and Wildlife Service.

Lennon, R. E., Hunn, J. B., Shnick, R. A. & Burress, R. M. (1971). Reclamation of ponds, lakes, and streams with fish toxicants: a review. *FAO Fisheries Technical paper*, 100, also Washington: US Dept. of the Interior, Fish & Wildlife Service.

Lennon, R. E. & Parker, P. S. (1959). The reclamation of Indian and Abrams Creeks in Great Smoky Mountains National Park. US Fish and Wildlife Service, Special Sci. Rep. 306, 1–22.

Leveque, C., Odei, M. & Thomas, M. P. (1977). The onchocerciasis control programme and the monitoring of its effect on the riverine biology of the Volta River Basin, In *Ecological Effects of Pesticides*, eds F. H. Perry & K. Mellanby, pp. 133–43. London: Academic Press.

Lewis, D. J. & Bennett, G. F. (1974). An artificial substrate for the quantitative comparison of the densities of larval simuliid (Diptera) populations. *Canadian Journal of Zoology*, 52(6), 773–5.

Lewis, D. J. & Bennett, G. F. (1975). The blackflies (Diptera: Simuliidae) of insular Newfoundland. III. Factors affecting the distribution and migration of larval simuliids in small streams on the Avalon Peninsula. *Canadian Journal of Zoology*, 33(2), 114–23.

Lichtenberg, J. J., Eichelberger, J. W., Dressman, R. C. & Longbottom, J. E. (1970). Pesticides in surface waters in the United States – a 5 year summary. *Pesticide Monitoring Journal*, 4, 71–86.

LOTEL (1977). The environmental impact of Sevin-4-oil (carbaryl) on a forest and aquatic ecosystem. *Lake Ontario Environ. Lab. (Oswego, N.Y.) Report*, 215.

Lowden, G. F., Saunders, C. L. & Edwards, R. W. (1969). Organochlorine insecticides in water. *Journal of the Society of Water Treatment and Examination*, 18, 275–96.

McAllister., Mauck, W. L. & Mayer, F. L. (1972). A simplified device for metering chemicals in intermittent-flow bioassays. *Transactions of the American Fisheries Society*, 10(3), 555–7.

McCullough, F. S., Gayral, P., Duncan, J. & Christie, J. D. (1980). Molluscicides in schistosomiasis control. *Bulletin of the World Health Organization*, 58(5), 681–9.

McCullough, F. S. & Mott, K. E. (1983). The role of molluscicides in schistosomiasis control. *World Health Organization Documentary Series*, WHO/VBC/83.879.

Macek, K. J., Buxton, K. S., Derr, S. K., Deane, J. E. & Sauter, S. (1976). Chronic toxicity of lindane to selected aquatic invertebrates and fish. *Service Rep.* No. EPA-600/3-76-046. Washington: US Environmental Protection Agency.

Macek, K. J. & Korn, S. (1970). Significance of the food chain – DDT accumulation by fish. *Journal of the Fisheries Research Board of Canada*, 27, 1496–8.

McIntyre, A. E. & Lester, J. N. (1982). Polychlorinated biphenyl and organichlorine insecticide concentrations in forty sewage sludges in England. *Environmental Pollution, B.*, 24, 225–30.

References

Maciorowski, H. D. & Clarke, R. M. (1980). Advantages and disadvantages of using invertebrates in toxicity testing, Aquatic Invertebrate Bioassays. *American Society for Testing and Materials, Tech. Publ.*, 715, pp. 36–47.

Mackey, A. P., Cooling, D. A. & Berrie, A. D. (1984). An evaluation of sampling strategies for qualitative surveys of macro-invertebrates in rivers using pond nets. *Journal of Applied Ecology*, 21, 515–34.

McLeese, D. W., Sargeant, D. B., Metcalfe, L. D., Zitko, V. & Burridge, L. E. (1980a). Uptake and excretion of aminocarb, nonyl phenol and pesticide diluent 585 by mussels (*Mytilus edulis*). *Bulletin of Environmental Contamination and Toxicology*, 575–81.

McLeese, D. W., Zitko, V., Metcalfe, C. D. & Sargeant, D. B. (1980b). Lethality of aminocarb and the components of the aminocarb formulation to Juvenile Atlantic salmon, marine invertebrates and a freshwater clam. *Chemosphere*, 9, 79–82.

McMahon, J. P., Highton, R. B. & Goiny, H. (1958). The eradication of *Simulium neavei* from Kenya. *Bulletin of the World Health Organization*, 19, 75–107.

Maki, A. W. (1980). Evaluation of toxicant effects on structure and function of model stream communities; correlations with natural stream effects. In *Microcosms in Ecological Research*, ed. J. P. Giesy, pp. 583–609. Washington: US Dept. Energy, Tech. Info. Centre.

Maki, A. W., Geissel, L. & Johnson, H. E. (1975). Comparative toxicity of larval lampricide (TFM: 3-trifluoromethyl-4-nitrophenol) to selected benthic macro-invertebrates. *Journal of the Fisheries Research Board of Canada*, 32, 1455–9.

Maki, A. W. & Johnson, H. E. (1976a). Evaluation of a toxicant on the metabolism of model stream communities. *Journal of the Fisheries Research Board of Canada*, 33, 2740–6.

Maki, A. W. & Johnson, H. E. (1976b). The freshwater mussel (Anodonta sp) as an indicator of environmental levels of 3-trifluoromethyl-4-nitrophenol (TFM). *Investigations in Fish Control*, No. 70. Washington: US Dept. of the Interior, Fish and Wildlife Service.

Maki, A. W. & Johnson, H. E. (1977). Kinetics of lampricide (TFM: 3-trifluoromethyl-4-nitrophenol) residues in model stream communities. *J. Fish. Res. Bd. Canada*, 34, 276–81.

Marker, A. F. H. & Casey, H. (1983). Experiments using an artificial stream to investigate the seasonal growth of chalk-stream algae. *Freshwater Biological Assn. Ann. Rep.*, pp. 63–75.

Marking, L. L. (1972). Salicylanilide L, an effective non-persistent candidate piscicide. *Transactions of the American Fisheries Society*, 101(3), 526–33.

Marking, L. L. & Chandler, J. H. (1978). Survival of two species of freshwater clams, *Corbicula leana* and *Magnonaias boykiniana*, after exposure to antimycin. US Dept. of the Interior, Fish and Wildlife Service, Washington. *Investigations in Fish Control*, No. 83, p. 5.

Mason, J. C. (1976). Evaluating a substrate tray for sampling the invertebrate fauna of small streams, with comments on general sampling problems. *Archives für Hydrobiologie*, 78(1), 51–70.

Matthiessen, P., Fox, P. J., Douthwaite, R. J. & Wood, A. B. (1982). Accumulation of endosulphan residues in fish and their predators after aerial spraying for the control of tsetse fly in Botswana. *Pesticide Science*, 13, 39–48.

Mauck, W. L., Olson, L. E. & Marking, L. L. (1976). Toxicity of natural pyrethrins and five pyrethroids to fish. *Archives of Environmental Contamination and Toxicology*, 4, 18–29.

References

Meier, P. G., Fook, D. C. & Lagler, K. F. (1983). Organochlorine pesticide residues in rice paddies in Malaysia, 1981. *Bulletin of Environmental Contamination and Toxicology*, 30, 351–7.

Meijers, A. P. & van der Leer, R. C. (1976). The occurrence of organic micropollutants in the river Rhine and the river Meuse in 1974. *Water Research*, 10, 597–604.

Merna, J. W. & Eisele, P. J. (1973). The effects of methoxychlor on aquatic biota. *USEPA. Ecol. Res. Service*, EPA-R3-73-046. Washington, DC: US Govt. Printing Office.

Metcalfe, R. L., Kapoor, I. P., Lu, Po-Yung, Schuth, C. K. & Sherman, P. (1973). Model ecosystem studies of the environmental fate of six organochlorine pesticides. *Environmental Health Perspectives*, June, 35–44.

Metcalfe, R. L. & Sanborn, J. R. (1975). Pesticides and environmental quality in Illinois. *Illinois Natural History Bulletin*, 31(9), 381–435.

Metcalfe, R. L., Sangha, G. K. & Kapoor, I. P. (1971). Model ecosystem for the evaluation of pesticide biodegradability and ecological magnification. *Environmental Science and Technology*, 5(8), 709–13.

Meyers, F. P. & Schnick, R. A. (1976). The approaching crisis in the registration of fishery chemicals. *Southeastern Association of Game and Fish Commissioners*, 30th Ann. Conf. pp. 5–14.

Meyers, F. P. & Schnick, R. A. (1983). Sea lamprey control techniques; past, present and future. *Journal of Great Lakes Research*, 9(3), 354–8.

Miles, J. R. W. (1976). Insecticide residues on stream sediments in Ontario, Canada. *Pesticide Monitoring Journal* 10, 87–91.

Miles, J. R. W. & Harris, C. R. (1971). Insecticide residues in a stream and a controlled drainage system in agricultural areas in southwestern Ontario, 1970. *Pesticide Monitoring Journal*, 5(3), 289–94.

Miles, J. R. W. & Harris, C. R. (1973). Organochlorine insecticide residues in streams draining agricultural, urban-agricultural, and resort areas of Ontario, Canada – 1971. *Pesticide Monitoring Journal*, 6(4), 363–8.

Ministry of Agriculture, Fisheries & Food (MAFF) (1985). *Guidelines for the Use of Herbicides In or Near Water Courses*. London: HMSO. MAFF booklet B2078.

Mohsen, Z. H. & Mulla, M. S. (1982). Field evaluation of *Simulium* larvicides on target and non-target insects. *Environmental Entomology*, 11(2), 390–8.

Molyneux, D. H., Baldry, D. A. T., van Wetters, P., Takken, W. & de Raadt, P. (1978). The experimental application of insecticides from a helicopter for the control of riverine populations of *Glossina tachinoides* in West Africa. I. Objectives, experimental area and insecticides evaluated. *Pesticides Abstracts and News Summary*, 24(4), 391–403.

Montreal Engineering Co. (1981). *Fenitrothion Accummulation by Plants and Invertebrates in Two Experimentally Sprayed Streams in New Brunswick*. Fredrichton, NB: Montreal Engineering Co.

Morrison, B. R. S. & Wells, D. E. (1981). The fate of fenitrothion in a stream environment and its effect on the fauna, following aerial spraying of a Scottish forest. *Science of the Total Environment*, 19, 233–52.

Mount, D. I. & Brungs, W. A. (1967). A simplified dosing apparatus for fish toxicity studies. *Water Research*, 1, 21–29.

Mount, D. I. & Warner, R. E. (1965). A serial dilution apparatus for continuous delivery of various concentrations of materials in water. *Publ. Hlth Ser. Publ.*, No. 999-WP-23. Washington: US Dept. Health, Education and Welfare.

References

Mount, M. E. & Oehme, F. W. (1981). Carbaryl; a literature review. *Residue Reviews*, 80, 1–64.

Muirhead-Thomson, R. C. (1957). Laboratory studies on the reactions of *Simulium damnosum* larvae to insecticides. *American Journal of Tropical Medicine*, 6, 920–34.

Muirhead-Thomson, R. C. (1969). A laboratory technique for establishing *Simulium* larvae in an experimental channel. *Bulletin of Entomological Research*, 59, 533–6.

Muirhead-Thomson, R. C. (1971). *Pesticides and Freshwater Fauna*. London & New York: Academic Press.

Muirhead-Thomson, R. C. (1973). Laboratory evaluation of pesticide impact on stream macroinvertebrates. *Freshwater Biology*, 3, 479–98.

Muirhead-Thomson, R. C. (1977). Comparative tolerance levels of blackfly (*Simulium*) larvae to permethrin (NRDC 143) and temephos. *Mosquito News*, 37(2), 172–9.

Muirhead-Thomson, R. C. (1978*a*). Lethal and behavioural impact of permethrin (NRDC 143) on selected stream macroinvertebrates. *Mosquito News*, 38(2), 185–90.

Muirhead-Thomson, R. C. (1978*b*). Lethal and behavioural impact of chlorpyrifos methyl and temephos on select stream macroinvertebrates; experimental studies on downstream drift. *Archives of Environmental Contamination and Toxicology*, 7, 139–47.

Muirhead-Thomson, R. C. (1978*c*). Relative susceptibility of stream macroinvertebrates to temephos and chlorpyrifos, determined in laboratory continuous-flow systems. *Archives of Environmental Contamination and Toxicology*, 7, 129–37.

Muirhead-Thomson, R. C. (1979). Experimental studies on macroinvertebrate predator-prey impact of pesticides. The reactions of *Rhyacophila* and *Hydropsyche* (Trichoptera) larvae to *Simulium* larvicides. *Canadian Journal of Zoology*, 57(11), 2264–70.

Muirhead-Thomson, R. C. (1981*a*). Relative toxicity of decamethrin, chlorphoxim and temephos (Abate) to *Simulium* larvae. *Tropenmedizin und Parasitologie*, 32, 189–93.

Muirhead-Thomson, R. C. (1981*b*). Tolerance levels of select stream macroinvertebrates to the *Simulium* larvicides, chlorphoxim and decamethrin. *Tropenmedizin und Parasitologie*, 32, 265–8.

Muirhead-Thomson, R. C. (1983). Time/concentration impact of the *Simulium* larvicide, Abate, and its relevance to practical control programmes. *Mosquito News*, 43(1), 73–6.

Mulla, M. S. & Darwazeh, H. A. (1976). Field evaluation of mosquito larvicides and their impact on some non-target insects. *Mosquito News*, 36(3), 251–6.

Mulla, M. S., Darwazeh, H. A. & Dhillon, M. S. (1980). New pyrethroids as mosquito larvicides and their effects on non-target organisms. *Mosquito News*, 40(1), 6–12.

Mulla, M. S., Darwazeh, H. A. & Majori, G. (1975). Field efficiency of some promising mosquito larvicides and their effects on non-target organisms. *Mosquito News*, 35(2), 179–85.

Mulla, M. S. & Mian, L. B. (1981). Biological and environmental impacts of the insecticides malathion and parathion on non-target biota in aquatic ecosystems. *Residue Reviews*, 78, 101–35.

Mulla, M. S., Mian, L. S. & Kawecki, J. A. (1981). Distribution, transport, and fate of the insecticides malathion and parathion in the environment. *Residue Reviews*, 81, 1–159.

Mulla, M. S., Navvab-Gojrati, H. A. & Darwazeh, H. A. (1978). Biological activity and

261

References

longevity of new synthetic pyrethroids against mosquitoes and some non-target insects. *Mosquito News*, **38**(1), 90–6.

Nash, R. W., Peterson, J. W. & Chansler, J. F. (eds) (1971). *Environmental Studies in the Use of Accothion for Spruce Budworm Control.* State of Maine: US Dept. of Agriculture & US Dept. of the Interior.

National Research Council Canada (1975). *Fenitrothion; the Effects of its Use on Environmental Quality and its Chemistry.* Associate Committee on scientific criteria for environmental quality. NRCC, No. 14104.

National Research Council Canada (1985). *TFM and Bayer 73: Lampricides in the Aquatic Environment.* NRCC No. 22488.

Newman, J. F. (1976). Assessment of the environment impact of pesticides. *Outlook on Agric.* **9**(1), 9–15.

Norris, L. A. (1981). The movement, persistence, and fate of the phenoxy herbicides and TCDD in the forest. *Residue Reviews*, **80**, 66–126.

Norris, L. A., Lorz, H. W. & Gregory, S. V. (1983). Influence of forest and rangeland management on anadromous fish habitat in western North America. Forest Chemicals. *US Dept. of Agric., Forest Service., Tech. Rep.*, PNW-149.

NTIS (1972). *Effects of Pesticides in Water.* A report to the States. Environmental Protection Agency. Nat. Tech. Info. Serv. US Dept of Commerce, PB-222 320.

NTIS (1975). *Methods for Acute Toxicity Tests with Fish, Macroinvertebrates, and Amphibians.* National Water Quality Laboratory, US Nat. Tech. Info. Serv. PB-242 105.

OCP (1983). *Onchocerciasis Control Programme in the Volta River Basin Area: Progress report of the World Health Organization for 1983.* Ouagadougou, Upper Volta: Joint Programme Committee. OCP/PR/83.2.

OCP (Onchocerciasis Control Programme) (1985a). *Report of activities, May–October 1984: Vector Control Unit.* No. 016/VCU/85, pp. 60.

OCP (1985b). *Revue Annuelle de la Recherche sur les Insecticides à OCP 19–20 Avril 1985 a Bouake (Côte d'Ivoire).* No. 421/VCU/ADM/3.4.

OCP (1985c). *Reunion Annuelle des Hydrobiologistes Chargés de la Surveillance Aquatique des Rivières de la Zone du Programme de Lutte Contre l'Onchocercose.* Ouagadougou. 12–15 Feb. 1985. No. 203/VCU/TEC/8.4.

Odum, W. E., Woodwell, G. M. & Wurster, C. F. (1969). DDT residues absorbed from organic detritus by fiddler crabs. *Science*, **164**, 576–7.

Opong-Mensah, K. (1984). A review of temephos with particular reference to the West African Onchocerciasis Control Programme. *Residue Reviews*, **91**, 47–69.

Park, P. O., Gledhill, J. A., Alsop, N. & Lee, C. W. (1972). A large-scale scheme for the eradication of *Glossina morsitans morsitans* Westw. in the western province of Zambia by aerial ultra-low-volume application of endosulphan. *Bulletin of Entomological Research*, **61**, 373–84.

Penney, G. H. (1971). Summary report of the effects of forest spraying in New Brunswick in 1972, on juvenile salmon and aquatic insects. Appendix 4 in *1971 Report of the Interdepartmental Committee on Forest Spraying Operations.* Ottawa.

Philippon, B., Escaffre, H. & Sechan, Y. (1976a). Control of *Simulium damnosum*, vector of human onchocerciasis in West Africa. V. Study of the insecticidal action of Abate EC 200 applied from aircraft by rapid release during dry seasons. *WHO, documentary series*, VBC/76.683.

Phillippon, B., Sechan, Y & Escaffre, H. (1976b). Control of *Simulium damnosum*, vector

262

of human onchocerciasis in West Africa. V. Study of the insecticidal action of Abate 200 EC applied from aircraft by rapid-release during dry seasons. *WHO Documentary series*, WNO/VBC/76.618.

Pinder, L. C. V. (1985). Studies on Chironomidae in experimental recirculating stream systems. I. *Orthocladius (Euorthocladius) calvus* sp.nov. *Freshwater Biology*, **15**, 235–41.

Poels, C. L. M. (1975). Continuous automatic monitoring of surface water with fish. *Water Treatment and Examination*, **24**(1), 46–56.

Quelennec, G., Dejoux, C., de Merona, B. & Miles, J. W. (1977). Chemical monitoring for temephos in mud, oysters and fish from a river within the Onchocerciasis Control Programme in the Volta River basin area. *WHO Documentary series*, VBC/77.683.

Quillevere, D., Bellec, C., LeBerre, R., Escaffre, H., Grebault, S., Kulzer, H., Quedraoga, J., Pendriez, B. & Philippon, B. (1976a). Control of *Simulium damnosum*, vector of human onchocerciasis in West Africa. IV. Evaluation by helicopter of new formulations and simulation of an insecticide operation. *WHO Documentary series*, WHO/VBC/76.617.

Quillevere, D., Duchateau, B., Escaffre, H., Grebault, S., Lee, C. W., Mouchet, J. & Pendriez, B. (1976b). Control of *Simulium damnosum*, vector of human onchocerciasis in West Africa. III. Application by aeroplane during the wet season. Methods of application – new formulations. A summary table of tests by classic and aerial application of new insecticides and new formulations. *WHO Documentary series*, VBC/76.616.

Raybould, J. N. & Grunewald, J. (1975). Present progress towards the colonization of African Simuliidae (Diptera). *Tropenmedizin und Parasitologie*, **26**, 155–68.

Richardson, G. M., Qadri, S. U. & Jessiman, B. (1983). Acute toxicity, uptake, and clearance of Aminocarb by the aquatic isopod Caecidolea racovitzai. 7, 552–7.

Roberts, J. R., Greenhalgh, R. & Marshall, W. K. (1977). (eds) Fenitrothion: the long term effects of its use in forest ecosystems. Proceedings of a Symposium. Nat. Res. Council Canada. NRCC/CNRC. pp. 628.

Robson, T. O. & Barrett, P. R. F. (1977). Review of effects of aquatic herbicides. In *Ecological Effects of Pesticides* ed. F. H. Perring & K. Mellanby pp. 111–18. *Linn. Soc. Sym. Series.*, No. 5. London: Academic Press.

Royal Commission (1979). *Royal Commission on Environmental Pollution*. 7th Report (Agriculture and Pollution). London: HMSO.

Ruber, E. & Kocor, R. (1976). The measurement of upstream migration in a laboratory stream as an index of potential side-effects of temephos and chlorpyrifos on *Gammarus fasciatus* (Amphipoda. Crustacea). *Mosquito News*, **36**(4), 424–9.

Russell-Smith, A. & Ruckert, E. (1981). The effects of aerial spraying of endosulphan for tsetse fly control on aquatic invertebrates in the Okavango swamps, Botswana. *Environmental Pollution*, A, **24**, 57–73.

Rye, R. P. & King, E. L. (1976). Acute toxic effects of two lampricides to twenty-one freshwater invertebrates. *Transactions of the American Fisheries Society*, **105**(2), 322–6.

Salford University (1981). *World Health Organization Onchocerciasis Control Programme: Preliminary Report on the Effect of Abate on Non-target Aquatic Fauna*, pp. B. 1–45. Geneva: World Health Organization.

Samman, J. & Thomas, M. P. (1978a). Effects of an organophosphorus insecticide,

References

Abate, used in the control of *Simulium damnosum* on non-target benthic fauna. *International Journal of Environmental Studies*, **12**(2), 141–4.

Samman, J. & Thomas, M. P. (1978*b*). Changes in zooplankton populations in the White Volta with particular reference to the effect of Abate. *International Journal of Environmental Studies*, **12**, 207–14.

Sanborn, J. R. & Ching-Chieh Yu (1974). The fate of dieldrin in a model ecosystem. *Bulletin of Environmental Contamination and Toxicology*, **10**(6), 340–6.

Sanders, H. O. (1969). Toxicity of pesticides to the crustacean Gammarus lacustris. *US Bureau of Sports Fisheries and Wildlife; Tech. Paper*, No. 25. Washington: US Bureau of Sports, Fisheries and Wildlife.

Sanders, H. O. (1977). Toxicity of the molluscicide Bayer 73 and residue dynamics of Bayer 2353 in aquatic invertebrates. *Investigations in Fish Control*, No. 78. Washington: US Dept. of the Interior.

Sanders, H. O. & Cope, O. B. (1966). Toxicity of several pesticides to two species of Cladocerans. *Transactions of the American Fisheries Society*, **95**, 165–9.

Sanders, H. O. & Cope, O. B. (1968). The relative toxicities of several pesticides to naiads of three species of stoneflies. *Limnology and Oceanography*, **13**(i), 112–17.

Sanders, H. O. & Walsh, D. F. (1975). Toxicity and residue dynamics of the lampricide 3-trifluoromethyl-4-nitrophenol (TFM) in aquatic invertebrates. *Investigations in Fish Control*, No. 59. Washington: US Dept. of the Interior, Fish and Wildlife Service.

Scherer, E. (ed.) (1979). *Toxicity Tests for Freshwater Organisms*. Govt of Canada Fisheries and Oceans (Special publication of fisheries and aquatic sciences, 44).

Scorgie, H. R. A. (1980). Ecological effects of the aquatic herbicide cyanatryn on a drainage channel. *Journal of Applied Ecology*, **17**, 207–25.

Sebastien, R. J. & Lockhart, W. L. (1981). The influence of formulation on toxicity and availability of a pesticide (methoxychlor) to blackfly larvae (Diptera: Simuliidae), some non-target aquatic insects and fish. *Canadian Entomologist*, **113**, 281–93.

Shires, S. W. (1983). The use of small enclosures to assess the toxic effects of cypermethrin on fish under field conditions. *Pesticide Science*, **14**, 475–80.

Shires, S. W. & Bennett, D. (1985). contamination and effects in freshwater ditches resulting from an aerial application of ocypermethrin. *Ecotoxicology and Environmental Safety*, **9**, 145–58.

Shriner, C. & Gregory, T. (1984). Use of artificial streams for toxicological research. *CRC Critical Reviews in Toxicology*, **13**(3), 253–81.

Sloof, W. (1983). Benthic macroinvertebrates and water quality assessment; some toxicological considerations. *Aquatic Toxicology*, **4**, 73–82.

Sloof, W. & Canton, J. H. (1983). comparison of the susceptibility of 11 freshwater species to 8 chemical compounds. II. (Semi)chronic toxicity tests. *Aquatic Toxicology*, **4**, 271–82.

Sloof, W., Canton, J. H. & Hermens, J. L. M. (1983*a*). Comparison of the susceptibility of 22 freshwater species to 15 chemical compounds. I. (Sub)acute toxicity tests. *Aquatic Toxicology*, **4**, 113–28.

Sloof, W., De Zwart, D. & Van de Kerkhoff, J. F. J. (1983*b*). Monitoring the Rivers Rhine and Meuse in the Netherlands for toxicity. *Aquatic Toxicology*, **4**, 189–98.

Smies, M., Evers, R. H. J., Peijnenburg, F. H. M. & Koeman, J. H. (1980). Environmental aspects of field trials with pyrethroids to eradicate tsetse fly in Nigeria, *Ecotoxicology and Environmental Safety*, **4**, 114–28.

Smith, A. J. (1966) Toxicity of the lamprey larvicide, 3-trifluoromethyl-4-nitrophenol to selected aquatic invertebrates. Hammond Bay Biological station, Millersburg, Michigan: US Bureau of Commercial Fisheries.

Smith, A. J. (1967). The effect of the lamprey larvicide 3-trifluoromethyl-4-nitrophenol, on selected aquatic invertebrates. *Transactions of American Fisheries Society* **96**(4), 410–13.

Smith, G. E. & Isom, B. G. (1967). Investigations on effects of large-scale applications of 2,4-D on aquatic fauna and water quality. *Pesticide Monitoring Journal*, **1**(3), 16–21.

Smith, R. L. & Hargreaves, B. R. (1983). A simple toxicity apparatus for continuous flow with small volumes: demonstration with mysids and naphthalene. *Bulletin of Environmental Contamination and Toxicology*, **30**, 406–11.

Sodergren, A. (1973). Transport, distribution, and degradation of chlorinated hydrocarbon residues in aquatic model ecosystems. *Oikos*, **24**, 30–41.

Sodergren, A., Svensson, B. & Ulfstrand, S. (1972). DDT and PCB in south Swedish streams. *Environmental Pollution*, **3**, 25–36.

Solbe, J. F. de L. G. (1979). Studies on the effects of pollution on freshwater fish. In *Biological Aspects of Freshwater Pollution*, ed. O. Ravera. Commission of European Committees. Pergamon Press, Oxford.

Solbe, J. F. de L. G. (1982). *Some responses of Fish to Pollutants*. Institute of Fisheries Management Fisheries Conference, Londonderry, & Water Research Centre, Stevenage, Herts, UK.

Sosiak, A. (1982). Caged fish and crayfish studies. In *Permethrin in New Brunswick Salmon Nursery Streams*, ed. P. D. Kingsbury, pp. 71–80. Sault Ste. Marie, Ontario: Forest Pest Management Institute.

Spehar, R. L., Anderson, R. L. & Fiandt, J. T. (1978). Toxicity and bioaccumulation of cadmium and lead in aquatic invertebrates. *Environ. Poll.*, **15**, 195–208.

Sprague, J. B. (1969) Measurement of pollutant toxicity to fish. I. Bioassay methods for acute toxicity. *Water Research*, **3**, 793–821.

Sprague, J. B. (1971). Measurement of pollutant toxicity to fish, III. Sublethal effects and 'safe' concentrations. *Water Research*, **5**, 245–66.

Stanley, J. G. & Trial, J. G. (1980). Disappearance constants of carbaryl from streams contaminated by forest spraying. *Bulletin of Environmental Contamination and Toxicology*, **25**, 771–6.

Stephenson, R. R. (1982). Aquatic toxicology of cypermethrin. I. Acute toxicity to some freshwater fish and invertebrates in laboratory tests. *Aquatic Toxicology*, 175–85.

Stiles, A. R. & Quelennec, G. (1977). Summary of field trials of candidate larvicides for control of *Simulium damnosum* in Africa, 1967–76. *WHO Documentary series*, VBC/77.667. Geneva: WHO

Stratton, K. G. (ed.). (1982–84). *Maine Cooperative Spruce Budworm Suppression Project*, *Environmental Monitoring Reports*, Augusta, Maine: Maine Forest Service, Dept. of Conservation.

Sundaram, K. M. S., Kingsbury, P. D. & Holmes, S. B. (1984). Fate of chemical insecticides in aquatic environments: forest spraying in Canada. In *ACS Symposial Series*, No. 238 (*Chemical and Biological Controls in Forestry*) ed. W. Y. Garner & J. Harvey pp. 253–76. Washington: American Chemical Society.

Surber, E. W. (1937). Rainbow trout and bottom fauna production in one mile of stream. *Transactions of the American Fisheries Society*, **66**, 193–202.

References

Sutcliffe, D. W. (1971). Regulation of water and some ions in Gammarids (Amphipoda) II. *Gammarus pulex* (L). *Journal of Experimental Biology*, **55**, 345–55.

Sutcliffe, D. W. & Carrick, T. R. (1973). Studies on mountain streams in the English Lake District. 1. pH, calcium and the distribution of invertebrates in the River Duddon. *Freshwater Biology*, **3**, 437–62.

Suzuki, T. (1983). *A Guidebook for Guatemalan Onchocerciasis (Robles Disease) with Special Reference to Vector Control*. Guatemala–Japan cooperative project on onchocerciasis research and control. Guatemala, Central America.

Symons, P. E. K. (1977*a*) Dispersal and toxicology of the insecticide fenitrothion: predicting hazards of forest spraying. *Residue Reviews*, **68**, 1–36.

Symons, P. E. K. (1977*b*). Assessing and predicting effects of the fenitrothion spray program on aquatic fauna. In *Fenitrothion: the Long-term Effects of its Use in Forest Ecosystems*, pp. 391–414. National Research Council, Canada, NRCC No. 22488.

Symons, P. E. K. & Harding, G. D. (1974). Biomass changes of stream fishes after forest spraying with the insecticide fenitrothion. *Fisheries Research Board of Canada*, 432.

Symons, P. E. K. & Metcalfe, J. L. (1978). Mortality, recovery, and survival of larval *Brachycentrus numerosus* (Trichoptera) after exposure to the insecticide fenitrothion. *Canadian Journal of Zoology*, **56**, 1284–90.

Takken, W., Balk, F., Jansen, R. C. & Koeman, J. H. (1978). The experimental application of insecticides from a helicopter for the control of riverine populations of Glossina tachinoides in West Africa. VI. Observations on side effects. **24**(4), 455–66.

Thayer, A. & Ruber, E. (1976). Previous feeding history as a factor in the effects of temephos and chlorpyrifos on migration of *Gammaurs fasciatus* (Amphipoda, Crustacea) *Mosquito News*, **36**(4), 429–32.

Tooby, T. E. (1981). Predicting the direct toxic effects of aquatic herbicides to non-target organisms. In *Proceedings of Aquatic Weeds and their Control*, pp. 265–74. Weltesbourne, Warwick: Association of Applied Biologists.

Torblaa, R. L. (1968). Effects of lamprey larvicides on invertebrates in streams. *US Fish and Wildlife Service, Special Sci. Rep., Fisheries*, No. 572.

Townsend, C. R. & Hildrew, A. G. (1976). Field experiments on the drifting, colonization and continuous re-distribution of stream benthos. *Journal of Animal Ecology*, **43**(3), 759–72.

Travis, B. V. (1949). Studies of mosquito and other biting insect problems in Alaska. *Journal of Economic Entomology*, **42**, 451–7.

Travis, B. V. & Schuchman, S. M. (1968). Tests (1967) with black fly larvicides. *Journal of Economical Entomology*, **61**, 843–5.

Travis, B. V. & Wilton, D. (1965). A progress report on simulated stream tests of blackfly larvicides. *Mosquito News*, **25**(2), 112–17.

Trial, J. G. (1978). Effects of Sevin 4 oil on aquatic insect communities; a continuation of 1976 studies. In *Environmental Monitoring Reports from Cooperative Spruce Budworm Control Project*, Maine 1976 and 1977, pp. 124–40, Augusta, Maine: Maine Forest Service, Dept. of Conservation.

Trial, J. G. (1979). The effect of Sevin-4-oil on aquatic insect communities of streams (1976–1978). In *Environmental Monitoring of Cooperative Spruce Budworm Control Projects*, Maine, 1978, pp. 6–22. Augusta, Maine: Maine Forest Service, Department of Conservation.

Trial, J. G. (1980). The effect of Sevin 4 Oil on aquatic insect communities in streams

References

(1976–1979). In *Environmental Monitoring Reports from 1979 Maine Cooperative Spruce Budworm Suppression Project*, ed. K. G. Stratton pp. 52–70. Augusta, Maine: Department of Conservation.

Trial, J. G. (1982a). The effectiveness of upstream refugia for promoting recolonization of *Plecoptera* killed by exposure to carbaryl. *Journal of Freshwater Ecology*, 1(6), 563–7.

Trial, J. G. (1982b). *The Ecological Significance of Reduced Populations of Leaf Shredding Plecoptera in Streams Sprayed with Carbaryl*. Final report to US Forest Service, Northeast area State and Private Forestry.

Trial, J. G. (1984). A study of leaf-processing disruption in streams within spruce budworm suppression project carbaryl spray blocks. *Environmental Monitoring Reports from the 1983 Maine Spruce Budworm Suppression Project*, ed. K. G. Stratton. pp. 1–34. Augusta, Maine: Maine Forest Service, Department of Conservation.

Trial, J. G. & Gibbs, K. E. (1978). *Effects of Orthene, Sevin-4-Oil and Dylox on Aquatic Insects Incidental to Attempts to Control Spruce Budworm in Maine, 1976*. Forest Insect and Disease Management, Dept. of Entomology, University of Maine: US Dept. of Agriculture.

Troubat, J,-J. (1981). Dispositif a gouttières multiples destiné à tester *in situ* la toxicité de insecticides vis-à-vis des invertébrés benthique. *Revue Hydrobiologique Tropicale*, 14(2), 149–52.

Van Windeguth, D. L. & Patterson, R. S. (1966). The effects of two organic phosphate insecticides on segments of the aquatic biota. *Mosquito News*, 26, 377–80.

Varty, I. W. (1976). Side effects of pest control projects on arthropods other than target species. In *Aerial Control of Forest Insects in Canada*, ed. M. L. Prebble, pp. 266–75. Ottawa: Canadian Forest Service Publication.

Wallace, J. B. & Brady, V. E. (1974). Residue levels of dieldrin in aquatic invertebrates and effect of prolonged exposure on populations. *Pesticide Monitoring Journal*, 5(3). 295–300.

Wallace, R. R. & Hynes, H. B. N. (1975). The catastrophic drift of stream insects after treatments with methoxychlor (1,1,1-trichloro-2,2-bis (p-methoxyphenyl) ethane). *Environmental Pollution*, 8, 255–68.

Wallace, J. B. & Hynes, H. B. N. (1981). The effects of chemical treatment against blackfly larvae on the fauna of running waters. In *Blackflies*, ed. M. Laird, pp. 237–58. London: Academic Press.

Wallace, J. B. & Merritt, R. W. (1980). Filterfeeding ecology of aquatic insects. *Annual Review of Entomology*, 25, 103–32.

Wallace, R. R., Hynes, H. B. N. & Kaushik, N. K. (1975) Laboratory experiments on facors affecting the activity of *Gammarus pseudolimnaeus* Bousfield. *Freshwater Biology*, 5, 533–46.

Wallace, R. R., Hynes, H. B. N. & Meritt, W. F. (1976) Laboratory and field experiments with methoxychlor as a larvicide for Simuliidae (Diptera). *Environmental Pollution*, 10, 251–69.

Wallace, R. R., Merritt, W. F. & West, A. S. (1973a) Dispersion and transport of Rhodamine B dye and methoxychlor in running water; a preliminary study. *Environmental Pollution*, 4, 11–18.

Wallace, R. R., West, A. S., Downe, A. E. R. & Hynes, H. B. N. (1973b). The effects of experimental blackfly (Diptera: Simuliidae) larviciding with Abate, Dursban and Methoxychlor on stream invertebrates. *Canadian Entomologist*, 105, 817–31.

References

Walsh, G. E., Miller, C. W. & Heitmuller, P. T. (1971). Uptake and effects of dichlobenil in a small pond. *Bulletin of Environmental Contamination and Toxicology*, 6(3), 279–98.

Ward, D. V. & Busch, D. A. (1976). Effects of temephos, an organophosphorus insecticide, on survival and escape behaviour of the marsh fiddler crab Uca pugnax. *Oikos*, 27, 331–5.

Warren, C. E. & Davis, G. E. (1971). Laboratory stream research; objectives, possibilities, and constraints. *Annual Review of Ecological Systems*, 2, 111–44.

Waters T. F. (1961). Standing crop and drift of stream bottom organisms. *Ecology*, 42, 532–7.

Waters, T. F. (1962). Diurnal periodicity in the drift of stream invertebrates. *Ecology*, 43, 316–20.

Waters, T. F. (1964). Recolonization of denuded stream bottom areas by drift. *Transactions of the American Fisheries Society*, 93(3), 311–15.

Waters, T. F. (1965). Interpretation of invertebrate drift in streams. *Ecology*, 46(3), 327–34.

Waters, T. F. (1972). The drift of stream insects. *Annual Review of Entomology*, 17, 253–72.

Wauchope, R. D. (1978). The pesticide content of surface water draining from agricultural fields – a review. *Journal of Environmental Quality*, 7, 459–72.

Way, J. M., Newman, J. F., Moore N. W. & Knaggs, F. W. (1971). Some ecological effects of the use of paraquat for the control of weeds in small lakes. *Journal of Applied Ecology*, 8, 509–32.

Wells, D. E. & Cowan, A. A. (1984). Fate and distribution of mothproofing agents dieldrin and Eulan WA New in Loch Leven, Kinross, 1964–79. *Environmental Pollution*, 7, 11–33.

Wells, D. E., Morrison, B. R. S. & Cowan, A. A. (1978). Chemical and biological observations in streams following aerial spraying of forests with fenitrothion. In *Control of the Pine Beauty Moth by Fenitrothion in Scotland*, eds. A. V. Holden & D. Bevan, pp. 103–33. Oxford: Forestry Commission, Nuffield Press.

Welton, J. S., Ladle, M. & Bass, J. A. S. (1982). Growth and production of *Ephemeroptera* larvae from an experimental recirculating stream. *Freshwater Biology*, 12, 103–22.

Whitford, L. A., Dillard, G. E. & Schumacher, G. J. (1964). An artificial stream apparatus for the study of lotic organisms. *Limnology and Oceanography*, 9, 598–9.

Wildish, D. J. & Lister, N. A. (1973). Biological effects of fenitrothion in the diet of brook trout. *Bulletin of Environmental Contamination and Toxicology*, 10(6), 333–9.

Wildish, D. J. & Phillips, R. L. (1972) Acute lethality of fenitrothion to freshwater aquatic insects. *Fisheries Research Board of Canada, Manuscript Rep. Ser.*, No. 1210.

Williams, D. D. & Hynes, H. B. N. (1974) The occurrence of benthos deep in the substratum of streams. *Freshwater Biology*, 4, 233–56.

Williams, E. H., Mather, E. L. & Carter, S. M. (1984). Toxicity of the herbicides endothal and diquat to benthic crustacea. *Bulletin of Environmental Contamination and Toxicology*, 31, 418–22.

Williams, T. R. & Obeng, L. (1962). A comparison of two methods of estimating changes in Simulium larvae populations, with a description of a new method. *Annals of Tropical Medicine and Parasitology*, 56, 359–61.

Willoughby, L. G. & Sutcliffe, D. W. (1976). Experiments on feeding and growth of the amphipod *Gammarus pulex* (L) related to its distribution in the River Duddon. *Freshwater Biology*, **6**, 577–86.

Wilson, D. C. & Bond, C. E. (1969) The effects of the herbicides Diquat and Dichlobenil (Casaron) on pond invertebrates. Part I. Acute toxicity. *Transactions of the American Fisheries Society*, **3**, 438–43.

Wilton, D. P. & Travis, B. V. (1965). An improved method for simulated stream tests of blackfly larvicides. *Mosquito News*, **25**(2), 118–22.

Wolfe, L. S. & Peterson, D. G. (1958). A new method to estimate levels of infestation of blackfly larvae. *Canadian Journal of Zoology*, **36**, 863–7.

Wood, G. (1983). Residue studies. In *Permethrin in New Brunswick Salmon Nursery Streams*, ed. P. D. Kingsbury, pp. 19–20. Sault Ste. Marie, Ontario: Forest Pest Management Institute.

Woodwell, G. M., Wurster, C. F. & Isaacson, P. A. (1967). DDT residues in an east coast estuary; a case of biological concentration of a persistent insecticide. *Science*, **156**, (3776), 821–4.

World Health Organization (WHO) (1981). *Onchocerciasis Control Programme: Preliminary Report on the Effect of Abate on Non-target Aquatic Fauna.* Salford University Team, Fish, pp. 12, Invertebrates.

World Health Organization (1984). Annual meeting of the hydrobiologists in charge of the aquatic monitoring of rivers in the Onchocerciasis Control Programme area. *WHO Documentary Series*, OCP/VCU/HYBIO/84.5.

World Health Organization (1985a). Onchocerciasis control programme, progress report, May–October 1984. *WHO Documentary Series*, 016/VCU/85.

World Health Organization (1985b). Annual meeting of the hydrobiologists in charge of the aquatic monitoring of rivers in the Onchocerciasis control Programme area. *WHO Documentary Series*, No. 203/VCU/TEC/8.4.

World Health Organization (1985c) Annual review of research on insecticides in the onchocerciasis control programme, April 1985. *WHO Documentary Series*, No. 42./VCU/ADM3.4.

Wright, J. F., Armitage, P. D., Furse, M. T. & Moss, D. (1985). The classification and prediction of macroinvertebrate communities in British rivers. *Freshwater Biol. Assn. Ann. Rep.*, 1985, 80–93.

Wuhrmann, K. (1964). River bacteriology and the role of bacteria in self-purification of rivers. In *Principles and Applications in Aquatic Microbiology*, ed. H. Heukelekian, N. C. Dondero, pp. 167–92. New York: Wiley.

Wuhrmann, K., Ruchti, J. & Eichelberger, E. (1966). Quantitative experiments on self-purification with pure organic compounds. 3rd International Conference on Water Pollutions Research, Munich, 1–17.

Yasuno, M., Fukushima, S., Hasegawa, J., Shioyama, F. & Hatakeyama, S. (1982a). Changes in the benthic fauna and flora after application of temephos to a stream on Mt Tsukuba. *Hydrobiologia*, **89**, 205–14.

Yasuno, M., Hasegawa, J., Iwakuma, T., Imamura, N. & Sugaya, Y. (1982b). Effects of temephos on chironomid and plankton populations in eel culture ponds. *Japanese Journal of Sanitary Zoology*, **33**, 207–12.

Yasuno, M., Hatakeyama, S. & Hasegawa, J. (1978). Toxicity of temephos to *Anisogammarus* sp. under the different temperature and exposure period. *Japanese Journal of Sanitary Zoology*, **29**, 365–6.

Yasuno, M., Okita, J., Nakamura, Y., Hatakeyama, S. & Kasuga, S. (1981a). Effects of fenitrothion on benthi fauna in small streams of Mt Tsukuba, Japan. *Japanese Journal of Ecology*, **31**, 237–45.

Yasuno, M., Okita, J. & Hatakeyama, S. (1982c). Effect of temephos on macrobenthos in a stream of Mt Tsukuba. *Japanese Journal of Ecology*, **32**, 29–38.

Yasuno, M., Shioyama, F. & Hasegawa, J. (1981b). Field experiment on susceptibility of macrobenthos in streams to temephos. *Japanese Journal of Sanitary Zoology*, **32**(3), 229–34.

Yasuno, M., Sugaya, Y. & Iwakuma, T. (1985). Effects of insecticides on the benthic community in a model stream. *Environmental Pollution*, **38**, 31–43.

Yule, W. N. & Tomlin, A. D. (1971). DDT in forest streams. *Bulletin of Environmental Contamination and Toxicology*, **5**(6), 479–88.

Zabik, M. J. (1969). The contribution of urban and agricultural pesticide use to the contamination of the Red Cedar River. *Michigan Inst. Water Res.* Project, No. A-012.

Zitko, V. & McLeese, D. W. (1980). Evaluation of hazards of pesticides used in forest spraying to the aquatic environment. *Canad. Tech. Rep. Fisheries & Aquatic Sciences*, No. 985.

Zitko, V., McLeese, D. W., Metcalfe, C. D. & Carson, W. G. (1979). Toxicity of permethrin, decamethrin and related pyrethroids to salmon and lobster. *Bulletin of Environmental Contamination and Toxicology*, **21**, 338–43.

INDEX

271

Index

272

Index

Glossina
  morsitans, 162–3
  pallidipes, 162
  palpalis, 162–4
  swynnertoni, 162
  tachinoides, 162–4
Gnathonemus pictus, 165
Guatemala (S. America), 77, 210–11
Gusathion, 86
guthion, 62
Gyrinus, 100, 165

Hastaperla, 186
Helicopsyche, 218
Heliosoma, 74
HEOD, active principle of dieldrin, 15
heptachlor, 49
  epoxide, 49
Heptagenia, 186
herbicides, 223–9
Hess sampler, 138
Hexagenia, 29, 30, 215–18, 241
Hyalella, 24, 54, 65, 95, 98–9, 132, 225–7
  azteca, 65, 99, 132, 227
Hydra, 214
Hydropsyche, 12, 35, 61, 64–5, 67, 72,
  80–2, 85, 91, 186, 196–7, 200, 202,
  209, 218–19, 242
  pellucidula, 64, 80–1, 218

Indiana (USA), 106
International Joint Commission (IJC), 15
Ischnura, 54, 215, 220
  verticalis, 215
Isogenus, 186
Isoperla, 95, 186, 208–9
isopod, 24
Ivory Coast (W. Africa), 167, 200

Japan, 205–8, 210

Kamoe Valley, 164–6
Kenya (E. Africa), 97
'kicking' technique, 187

laboratory data, 92–3, 98–104
Lakes
  Chew (Somerset), 11
  Great (N. America), 14, 92, 97, 106, 113,
    213, 216–18, 223, 238
  Leven (Scotland), 239
  Michigan (N. America), 14, 17, 105, 215
  Muskoko (Canada), 15
  Nasser (Egypt), 16
  Neagh (N. Ireland), 13
  Nubia (Sudan), 16
  Ontario (N. America), 219
  Superior (N. America), 15, 215
lamprey, 94, 215, 222, 232

Lamsilis radiata, 106–7
LC (lethal concentration) definition, 23
Lebaycid, 63
Lebistes reticulatus, 116
leech, 27, 54, 214, 219
Leptocera, 224
Leptophos, 116
lethal concentration (LC) definition, 23
Leuctra, 96, 132, 135–6, 152, 155
Libellula, 226
lindane (hexachlorocyclohexane), 11–14,
  16, 18, 74–5, 80
Lumbriculus, 218
Lymnaea
  elodes, 75–6
  peregra, 228

MAC, definition, 214
Macrobrachium
  raridus, 164, 166
  vollenhovenii, 168
Macronema, 199, 202
Maine (USA), 98, 110, 154–8 passim, 238–9
malathion, 27, 62, 115
maple, 16
Mariotte bottle, 135
marsh fiddler crab, 90
Matacil, 111, 159, 238
may-fly, 14, 25, 29, 30, 52, 73, 83, 98,
  107, 109, 132, 135, 138, 142–3,
  146–8 152, 155–6, 165–8, 171, 178,
  185–6, 202–4, 214–18, 220, 226–7,
  241
median tolerance limit (TL$_m$), 29
Melosira, 95
methoxychlor, 4, 16, 71–2, 78–80, 88–9,
  105–7, 109, 115, 125, 176–80, 182,
  184–7
Michigan (USA), 52–3, 106
midge, 9, 26, 95, 97, 141, 144, 215–16,
  220, 226
Minnesota (USA), 106
MLC, definition, 214
Moina micrura, 104
mosquito, 9, 14, 66, 75, 90, 97, 114–16
mosquito fish, 114–16
Mount Tsukuba (Japan), 206–8
mussel, 106
Myriophyllum, 228

naphtha (heavy aromatic) HAN, 179
Nashwaak Experimental Watershed Project,
  New Brunswick, 129
Nemoura, 95–6, 135, 208–9
New Brunswick (Canada), 17–18, 87, 93,
  110–11, 123, 128–9, 131–3, 151,
  153–8 passim
Newfoundland, 128
niclosamide, 26

273